LAND APPLICATION
of
SEWAGE SLUDGE
and
BIOSOLIDS

LAND APPLICATION
of
SEWAGE SLUDGE
and
BIOSOLIDS

Eliot Epstein, Ph.D.

CRC Press
Taylor & Francis Group
Boca Raton   London   New York

CRC Press is an imprint of the
Taylor & Francis Group, an **informa** business

First published 2003 by Lewis Publishers

Published 2019 by CRC Press
Taylor & Francis Group
6000 Broken Sound Parkway NW, Suite 300
Boca Raton, FL 33487-2742

© 2003 by Taylor & Francis Group, LLC
CRC Press is an imprint of Taylor & Francis Group, an Informa business

First issued in paperback 2019

No claim to original U.S. Government works

ISBN-13: 978-0-367-45474-6 (pbk)
ISBN-13: 978-1-56670-624-7 (hbk)

Visit the Taylor & Francis Web site at
http://www.taylorandfrancis.com

and the CRC Press Web site at
http://www.crcpress.com

Library of Congress Cataloging-in-Publication Data

Epstein, Eliot, 1929–
    Land application of sewage sludge and biosolids / Eliot Epstein.
        p.   cm.
    ISBN 1-56670-624-6 (alk. paper)
    1. Land treatment of wastewater. 2. Sewage disposal in the
    ground--Environmental aspects. 3. Sewage sludge--Management. I. Title.
TD774 .E64 2002
628.3′8—dc21

2002073030
CIP

Library of Congress Card Number 2002073030

Preface

As indicated in Chapter 1 of this book, land application of sewage sludge and biosolids has been practiced for centuries. Over the past 40 years, I have been involved in various aspects of organic matter and soils. Since 1972, I have researched, studied, and published on various aspects of biosolids management, concentrating on composting and the public health aspects of land application. I have been an active member on the Water Environment Federation Residuals Committee and The U.S. Composting Council Board of Directors. In 1977, I published a book titled *The Science of Composting*.

It was very difficult to write the present book because the literature on this subject is enormous. I reviewed more than 4000 references during the year that it took me to write it, and cited more than 570 references. The interest and concerns of scientists throughout the world not only indicate the importance of the subject, but also show the dedication to disseminating information related to health and the environment. It is very evident that the majority of the scientific community think that the management and use of biosolids in a sound manner is proper and environmentally safe.

My objective in writing this book was to provide the reader with insights into the scientific writings and findings. Those who are interested in more detail can delve into the cited works to obtain more information.

One of the major aspects of land application of biosolids is the issue of risk to humans, animals, plants and the environment. The two major categories of risk are (1) voluntary and (2) imposed. People who smoke cigarettes do so voluntarily, even though they are aware of the risk. The same applies to driving a car or flying. These are all voluntary risks. However, we often do not have choices, but have a risk imposed upon us. The placement of nuclear power plants near communities is such an imposed risk. The most important issue related to imposed risks is how serious or great they are. We do not live in a risk-free society. The risk and vulnerability clearly impacted all of us on September 11, 2001. Consider the risk of using biosolids on human health and the environment in light of the following:

- The risk from contaminated food by *E. coli* O157:H7, which has resulted in numerous deaths and contamination of food by *Salmonella* sp. and other bacteria.
- The risk from contaminated water, which has resulted in many persons being ill, as well as numerous deaths.
- The risk from bacteria and viruses when using home toilets (Gerba, C.P., C. Wallis, and J.L. Melnick, 1975. Microbial hazards of household toilets: Droplet production and the fate of residual organism. *Applied Microbiology* 30(2):229-237).
- The risk of ingesting fish contaminated with mercury, as compared with ingesting mercury from biosolids-contaminated food.
- The risk of indoor air pollution from volatile organic compounds in carpets, paints, household cleaners, etc.
- The risk of bioaerosols when walking through the park or visiting a farm, as compared with the potential risk from bioaerosols from an outdoor composting facility.
- The risk of a Staphylococcus infection or infection by *Aspergillus fumigatus* in a hospital.

It is important to put the use of biosolids into the proper risk perspective. Biosolids can be disposed of only in the soil, water or air. The greatest advantages to land application are that we can control our activities and manage them. Once a contaminant is in the water or air, it is very difficult to control it. Our biosolids control and management practices need to be consistent with good scientific knowledge and judgment. It is the duty of the scientist to seek out the truth and to provide the engineer and biosolids manager with the knowledge and direction of how best to protect the health of humans, animals, plants and the environment. I hope this book will shed some light on the degree of potential risk from applying biosolids to land.

I would like to repeat my quotation from Berth Damon cited in my earlier book. It is also appropriate for land application of biosolids, as the organic matter will eventually turn into humus.

> To consider *humus* is to get a hint of the oneness of the universe. All flesh is grass, in more than the figurative sense the prophet intended. During the long history of this planet, weather has disintegrated rock, tiny lichens have made a speck of vegetable mold, countless generations of short-lived weeds have waxed fat for summer, giant forests have flourished for an aeon, and all in turn have died and given back to the earth more goodness than they have taken from it. All have been composted into humus. And the life of insects and of animals and of men which was sustained upon the life of these plants and upon the life of other animals, all these creatures too have enriched the surface of the earth with their excreta and finally with their bodies. All in turn have been composted into humus.

Acknowledgments

I am most grateful to Dr. Albert Page from the University of California, Riverside, a well-recognized soil scientist who has written numerous articles and books on this subject, who reviewed and commented on the chapters on heavy metals. Numerous individuals have provided me with insight and knowledge over the years. Several who coauthored articles with me include Drs. Rufus Chaney, Bob Dowdy, Terry Logan, Chuck Henry, Pat Millner and John Walker. All have contributed to our knowledge and understanding of the use and management of biosolids and organic matter. I would also like to thank Laure MacGibbon, our administrative assistant at Tetra Tech, Inc., for assisting me in editing this book.

My greatest thanks and appreciation go to my family. They have provided me with inspiration, support and encouragement throughout the entire process. My wife, Esther, deserves special thanks for the considerable time she spent proofreading the manuscript, contacting firms for permission to use material, and many other tasks throughout the process.

Author

Eliot Epstein is Chief Environmental Scientist for Tetra Tech, Inc. and an adjunct professor of public health at Boston University School of Public Health at the School of Medicine. He received his B.S. degree in Forestry from New York College of Forestry at Syracuse University, an M.S. degree in Agronomy from the University of Massachusetts, and a Ph.D. in soil physics from Purdue University. For 16 years he was a research leader for the U.S. Department of Agriculture's Agricultural Research Service and an adjunct professor of soil physics at the University of Maine. His research there concentrated on soil erosion and runoff and soil water relations of plants.

In 1972, Dr. Epstein transferred to the USDA ARS research center in Beltsville, Maryland, where he conducted research on the use of biosolids, and where, in 1975, he researched and developed the aerated static pile method (ASP). In 1980, he became president of E&A Environmental Consultants, Inc., a premier company in composting and beneficial use of organic materials. In that capacity, he was the principal-in-charge of numerous projects conducted by the staff located in Massachusetts, North Carolina, and Washington State.

Dr. Epstein has more than 30 years of experience in biosolids composting, and has managed or directed more than 400 composting projects in the United States, Canada and Europe. He consulted on composting and biosolids management for the US EPA, World Bank and United Nations. In 2001, Dr. Epstein and his staff joined Tetra Tech, Inc., a leading company in water reuse, wastewater and beneficial use of organic residues.

Table of Contents

CHAPTER **1**

Land Application of Biosolids: A Prospective

INTRODUCTION

The United States Environmental Protection Agency (USEPA) accepts the term *biosolid* for sewage sludge that is treated to meet the regulations in 40CFR503 (USEPA, 1994). Moreover, the term *sludge* is used for many different materials, both inorganic and organic. A better term would be sewage biosolids, because it indicates the biological nature of the residual and its origin. Many organic residuals, e.g., paper mill sludges, are biological and could be classified as biosolids. In this text, biosolids will be used to designate treated sewage sludge that meets USEPA Class A or B. Raw sludge, e.g., primary, waste-activated, secondary sludge, and those materials that do not meet the USEPA Class A or B will still be designated as sewage sludge.

Biosolids are derived from the treatment of wastewater. The wastewater, primarily derived from domestic sources, is treated at a publicly owned treatment works (POTW). This wastewater will also contain discharges from commercial and industrial enterprises. Many of these enterprises conduct pretreatment prior to discharging wastes into the sewer system. As a result of pretreatment, which was enforced by the Clean Water Act of 1974, biosolids have become less contaminated with trace elements that include heavy metals and organic compounds.

The predominant wastewater treatment processes are primary and secondary treatment. Tertiary treatment may be necessary before discharge of the clean effluent is allowed into certain classes of water bodies. Both the nature of the domestic waste and the secondary treatment, which is a biological treatment, result in sludge being predominantly organic matter.

Solids in the wastewater stream are removed during primary and secondary treatment. This sewage sludge, if it does not undergo further treatment, is often referred to as raw sludge. It is usually incinerated, landfilled or further treated. Further treatment may consist of digestion, composting or alkaline stabilization.

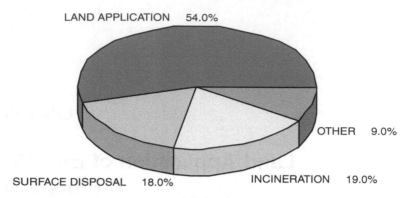

Figure 1.1 Use and disposal of sewage sludge and biosolids in the United States (USEPA, 1993).

After treatment, this material is called biosolids. Biosolids will contain inorganic material, plant nutrients, trace elements, organic compounds, and pathogens. (More data on characteristics of biosolids are presented in Chapter 2.) The organic nature of biosolids, along with plant nutrients and several trace elements, which are micronutrients for plants, makes it a valuable resource for land application. However, high concentrations of several trace elements, toxic organic compounds, and pathogens can preclude the beneficial use of biosolids.

USE AND DISPOSAL OF SEWAGE SLUDGE AND BIOSOLIDS

The USEPA in 1998 estimated that approximately 6,232,880 metric tons (tonnes) or 6,856,168 dry tons are generated annually in the United States. In its September 1999 report entitled Biosolids: Generation, Use, and Disposal in the U.S., the agency estimated that 7.1 million tons would be generated in 2000 and 8.2 million tons by 2010. The distributions for the most common methods of disposal or utilization are shown in percentage terms in Figure 1.1. Distribution and marketing, which include composting and heat drying, were estimated at approximately 3%. Ocean disposal has been discontinued since 1992. Because disposal to landfills has been severely curtailed in many states, land application has increased. Incineration is also decreasing because many incinerators cannot meet the Clean Air Act regulations or the USEPA 503 regulations.

In the United States there are vast differences in the way biosolids or sewage sludge is utilized or disposed. These differences, compiled by Bastian (1997), show that in New England and the Northeast less than 30% is land applied (Table 1.1).

A significant amount of biosolids from the Northeast is shipped out of state for either land application or landfills. Nearly 90% of the biosolids produced in the Northwest are land applied.

Table 1.2 compares the use and disposal of sludge and biosolids in 1989 and 1997. In that 9-year timeframe, land application that includes composting, heat drying, and other products that are distributed or marketed increased by 12%, landfilling and surface disposal decreased, incineration increased, and ocean disposal ceased.

Table 1.1 Biosolids/Sewage Sludge Utilized or Disposed by USEPA Regions in the United States

Region	Population served by POTWs[1]	Total biosolids/sludge production (dmt/yr)	Percent used/disposed by:			
			Land application[2]	Surface disposal[3]	Incineration	Other
I. New England	8,037,311	367,430	24	46	30	–
II. Northeast	21,726,101	605,046	30	14	23	33[4]
III. Mid Atlantic	18,152,556	1,040,206[5,6]	74	16	10	<1
IV. Southeast	24,510,111	1,050,325[5]	57	30	12	1
V. North Central	35,587,804	1,705,316	51	2	30	17[7]
VI. South Central	21,150,172	425,203[5]	53	45	2	–
VII. Midwest	9,036,498	511,712[5]	65.5	4	25.5	5
VIII. Rocky Mountains	6,262,873	111,880[5]	68	29	0	3[7]
IX. Southwest	30,432,899	819,050	51	36	4	9[7]
X. Northwest	5,634,539	220,000	89	2	9	–
U.S. Totals	180,530,874	6,856,168[6]	54	18	19	9

[1] Based on 1992 Needs Survey Data (USEPA, 1993).
[2] Includes all forms of land application practices, such as application of liquid, dewatered cake, dried, composted, alkaline stabilized, or otherwise processed product to cropland, forests, reclamation sites, lawns, parkland, etc.; use as organic fertilizer or soil amendment; use in potting mixes and the production of topsoil, etc. (including use as a daily or final landfill cover as land application).
[3] Includes co-landfilling with solid waste; monofilling; permanent disposal in piles or lagoons, etc.
[4] Shipped out of state for use/disposal (mainly by land application and landfill).
[5] Includes estimates based upon population served by POTWs and regional conditions.
[6] Some estimates (based upon the number of households served by POTWs) of total production for Pennsylvania would increase this number by 1,578,639 dmt/yr.
[7] Long-term storage.

Source: Bastian, 1997, Eur. Water Pollut. Control 7(2): 62–78. With permission.

Table 1.2   A Comparison of Sludge and Biosolids' Use or Disposal in The United States in 1989 and 1997

Use or Disposal	1989*	1997**
Land application & distribution and marketing	42	54
Landfill, monofill and surface disposal	24	18
Incineration	11	19
Ocean	3	0
Other	12	9

* USEPA analytical survey (USEPA, 1990).
** Bastian, 1997.

At the same time, the quality of biosolids has improved. Examples of some reductions of heavy metals at several wastewater treatment plants are shown in Table 1.3. (More detail on the characteristics of sludge and biosolids is provided in Chapter 2.)

Bastian (1997) reported that biosolids production in the United States has constantly increased. Over the past 10 to 20 years, not only have there have been major changes in sewage sludge/biosolids management practices, but also significant changes in equipment (Bastian, 1997). These include:

1. Expanded use of mechanical dewatering equipment and polymers.
2. A shift from the use of multiple hearth incinerators to fluid bed incinerators.
3. Less long-term lagoon storage.
4. Less use of landfilling with municipal solid waste, but an increase in use as a daily or final landfill cover material.
5. Increase in use of land application.
6. Increase in use of heat drying, pelletizing, composting, and alkaline stabilization processing of biosolids to generate products.
7. A phase-out of ocean dumping.
8. More long-distance transport of biosolids to end use/disposal sites.
9. Increase in energy recovery from incinerators, landfills, and anaerobic digesters.
10. Increase of contracted-out biosolids management operations.
11. Increase in multiple biosolids use and disposal practices, especially by the large authorities.

Bastian (1997) indicates that several factors influenced these trends in sewage sludge/biosolids processing and use/disposal practices:

1. Dramatic increases in biosolids production.
2. Improvement in quality of biosolids as a result of industrial pretreatment.
3. Improvements in biosolids handling equipment and processing.
4. Improvement in biosolids management practices.
5. Increase in costs for urban land, labor, and energy.
6. A willingness of management agencies to pay dramatically higher prices for biosolids processing, beneficial use, and disposal.
7. More information from research and development and monitoring activities.
8. Changes in regulatory policies and requirements.
9. Increase in public sensitivity on environmental and health issues and negative public reactions to biosolids processing sites and disposal facilities.

Table 1.3   Examples of Several POTWs Demonstrating Reductions in Biosolids' Heavy Metals

POTW (actual flow)	Source Years	Reported Reductions in Loadings to Biosolids
Bowling Green, KY (5 mgd)	1985–1988	Zn – 97%; Pb – 90% Cr – 72%; Ni – 100% Cd – 91%; Cu – 88%
Pocatello, ID	1985–1988	Zn – 45%; Pb – 36% Cr – 67%; Ni – 56% Cd – 57%; Cu – 43%
Albany, NY (7 mgd)	1983–1987	Zn – 94%; Pb – 98% Cr – 99%; Ni – 99% Cd - <50%; Cu – 99%
Springfield, OH (19.9 mgd)	1984–1989	Zn – 77%; Pb – 87% Cr – 79%; Ni – 50% Cd – 79%; Cu – 53%
Oswego, IL (22.8 mgd)	1985–1990	Zn – 56%; Pb – 47% Cr – 92%; Ni – 78% Cd – 96%; Cu – 50%
Louisville and Jefferson County, KY (97.9 mgd)	1982–1989	Zn – 79%; Pb – 68% Cr – 74%; Ni – 25% Cd – 83%; Cu – 54%
Metro Seattle, WA (156 mgd)	1981–1989	Pb – 46%; Cd – 38% Cu – 56%

Source: Bastian, 1997, *Eur. Water Pollut. Control* 7(2): 62–78. With permission.

More recent factors that are influencing biosolids management practices are land application bans, incinerator air emissions, and odors from composting operations.

Matthews (1999) reported on management methods in Europe. Table 1.4 shows his data. Disposal methods vary considerably in the different European countries. Matthews indicates that sludge/biosolids production in the European Union will have doubled from 1992 to 2002. Marine disposal will be discontinued, putting more pressure on land-based methods of utilization and disposal. In Greece, for example, 90% of the biosolids was landfilled. On average in Europe, 36.4% of the biosolids was used in agriculture; 41.6% were landfilled; and 10.9% was incinerated. The remaining 11.19% was dumped into the ocean or surface waters, or used for other purposes such forestry or recultivation.

SYSTEMS FOR THE USE OR DISPOSAL OF SEWAGE SLUDGE AND BIOSOLIDS

There are several options for the disposal or utilization of biosolids:

• Direct land application
• Composting and land application

Table 1.4. Biosolids Production and Disposal Methods in Europe

Country	Quantity 1000 dry tonnes/ year	Agriculture	Landfill	Incineration	Sea	Other (e.g., recultivation, forestry)
				Percent		
Austria	170	18	35	34	–	13
Belgium	59.2	29	55	15	–	1
Denmark	170.3	54	54	40.9	–	2
Finland	150	25	75	–	–	–
France	865.4	58	27	15	–	–
Germany	2681.2	27	54	14	–	5
Greece	48.2	10	90	–	–	–
Ireland	36.7	12	45	–	35	8
Italy	816	33	55	2	–	10
Luxembourg	8	12	88	–	–	–
Netherlands	335	26	51	3	–	20
Norway	95	58	44	–	–	–
Portugal	25	11	29	–	2	58*
Spain	350	50	35	5	10	–
Sweden	200	40	60	–	–	–
Switzerland	270	45	30	25	–	–
UK	1107	44	8	7	30	11
Total	7387	36.4	41.6	10.9	5.19	6

* Surface waters

Source: Data from Matthews, 1999.

- Heat drying and land application
- Incineration
- Landfilling
- Dedicated land disposal

Direct land application can be beneficially used in agriculture, forestry, and land reclamation. The biosolids can be applied in either in a liquid form with low solids or as a semisolid following dewatering. Because of the nature of the material, in that it contains a large amount of water, it is often applied within relatively short distances. Exceptions include New York City, where dewatered, digested biosolids are rail hauled and applied as far as Texas and Colorado. Similarly, cities in southern California have some of their biosolids trucked to Arizona. Chicago Greater Metropolitan District barged liquid biosolids to Fulton County, a distance of approximately 200 miles, and then applied the liquid biosolids to agricultural land.

Direct land application for beneficial use involves some form of partial stabilization such as digestion or alkaline stabilization. Digestion results in USEPA 503 Class B biosolid whereas alkaline stabilization can result in a Class A or B product. The USEPA 503 regulations allow for either a Class A or Class B to be applied to land (see Chapter 12 for more information about regulations).

Application of biosolids for beneficial uses considers the crop requirement for the plant nutrients as well as the accumulation of trace elements. When sewage sludge or biosolids are land applied as a dedicated land disposal method, no crop is grown and the amounts applied are usually greater than when the biosolids are beneficially used.

Composting and heat drying are forms of land application. These technologies produce a USEPA 503 Class A product that provides for a wider range of uses. Today, most of these products meet the "Exceptional Quality" criteria for trace elements. The drier nature of these two materials provides for bagging and shipping to greater distances. Bagged composted animal manure has been shipped from California to Germany and heat-dried products are shipped from Massachusetts to Florida.

Biosolids compost is primarily used as a soil amendment and is valued for the organic matter content (Epstein, 1997). Its primary use today is in the horticultural field. However, recently farmers in several areas of the United States have been using compost to improve their soil's physical properties such as water holding capacity and soil structure. The nitrogen content of compost is lower than heat-dried biosolids. Therefore, larger quantities can be applied to improve the soil's physical properties. Heat-dried biosolids are considered as a fertilizer and applied at fertilizer rates. They may be used as a supplement to other inorganic fertilizer material to increase the plant nutrient content. Milorganite is an example of a heat-dried biosolid product from the city of Milwaukee, Wisconsin. It is widely used as a fertilizer in the turf industry.

HISTORY OF LAND APPLICATION OF SEWAGE SLUDGE AND BIOSOLIDS

References to land application of wastes can be found in the Bible (Deuteronomy 23:13 and Judges 3:20). The Greek and Romans understood the benefits of basic sanitation to society. In Ephesus, Turkey, steam-heated toilets were in place and wastes were flushed into the Aegean Sea. The author, in a visit to Ephesus, noted this continuous flushing system using heated marble toilets. Doxiadis (1973) also indicated that domestic wastes were conveyed to land application areas. In the 16th century, effluents were being used for crop production in Bunzlau, Germany. This practice began in 1550 and continued for 300 years (De Turk, 1935).

Jewell and Seabrook (1979) cited Gerhard (1909), who indicated that the earliest documented sewage farm or sewage irrigation system was in Bunzlau in 1531. In England, in 1859, a Royal Commission on Sewage Disposal recommended the application of town sewage to land as a means of avoiding river pollution (Webber and Hillard, 1974). Another early sewage farm was established in Edinburgh, Scotland around 1860. Major cholera epidemics in London in the mid-1800s that claimed more than 25,000 lives stimulated the treatment of sanitation facilities (Jewell and Seabrook, 1979). By 1876, 35 towns in Britain used land treatment. Large sewage farms were established in Paris, France (1869) and in Berlin, Germany (1874).

Between 1870 and 1880 there were 30 to 50 land application sites. Land application was just beginning in the United States (Jewell and Seabrook, 1979). The first sewage crop irrigation system in the United States was established in Augusta, Maine in 1872. By 1900 there were more than 10 places in the United States using

Table 1.5 Some Early Land Application Systems

Location	Date started	Type of System	Area-ha	Flow m³/d
International				
Croydon-Bedlington, England	1860	Sewage farm	255	21,198
Paris, France	1869	Slow rate infiltration	6,475	299,050
Leamington, England	1870	Sewage farm	162	3,785
Berlin, Germany	1874	Sewage farm	27,519	—
Worclaw	1882	Sewage farm	809	105,992
Melbourne, Australia	1893	Slow rate infiltration	4,209	189,272
		Overland flow	1,416	264,981
Braunschweig, Germany	1896	Sewage farm	4,452	60,567
Mexico City, Mexico	1900	Slow rate infiltration	45,360	215,770
United States				
Calumet City, Michigan	1886	Rapid infiltration	4.86	4,543
Woodland, California	1889	Slow rate infiltration	97	15,899
Fresno, California	1891	Slow rate infiltration	1620	98,421
San Antonio, Texas	1895	Slow rate infiltration	1620	75,709
Vineland, New Jersey	1901	Rapid infiltration	5.67	3,028
Ely, Nevada	1908	Slow rate infiltration	567	5,678

Source: Data from Fuller, 1983.

land treatment and by 1980 more than 4,000. Table 1.5 shows some selected land applications systems in Europe and the United States.

Anderson (1959) stated that the use of sewage sludge as a fertilizer dates essentially from 1927, when a large activated sludge treatment plant went into operation for the city of Milwaukee, Wisconsin. Table 1.6 shows the utilization of biosolids as fertilizer from 1930 to 1957 (Anderson, 1959). The author indicated that for sanitary reasons, activated sludge must be heat-treated before it is sold as fertilizer. He stated that digested biosolids should be heat-treated or incorporated into the soil for several months before application on vegetables.

Table 1.6 Utilization of Biosolids as Fertilizer from 1930 to 1957

Year	Dried Activated Biosolids		Dried Digested Biosolids	
	As Separate Material	As Mixed Fertilizer	As Separate Material	As Mixed Fertilizer
1930	14,852	17,811	—	—
1940	26,307	78,992	23,000	3,000
1945	29,261	61,306	30,274	5,791
1950	63,366	55,965	30,738	6,581
1953	69,494	—	36,719	—
1957	93,152	—	37,577	—

Source: Adapted from Anderson, 1959.

Table 1.7   A Summary of the Major Wastewater Processes

Wastewater Treatment Level	Types of Sewage Sludge or Biosolids Produced

Screening and Grit Removal

Wastewater screening removes coarse solids that can interfere with mechanical equipment. Grit removal separates heavy inorganic, sandlike solids that would settle in channels and interfere with the treatment processes.	Screening and grit are handled as a solid waste and nearly always landfilled. This material is not considered sewage sludge or biosolids and is not governed by USEPA 40CFR503 regulations.

Primary Wastewater Treatment

Usually involves gravity sedimentation of screened, degritted wastewater to remove suspended solids prior to secondary treatment.	Sewage sludge by primary wastewater treatment usually contains 3%–7% solids; generally their water content can be easily reduce by thickening or dewatering.

Secondary Wastewater Treatment

Generally relies on biological treatment (e.g., suspended growth or fixed growth systems), in which microorganisms are used to reduce biochemical oxygen demand and remove suspended solids. Secondary treatment is the minimum treatment level required for POTWs under the Clean Water Act.	Sewage sludge produced by secondary wastewater treatment usually has a low solids content (0.5% –2%) and is more difficult to thicken and dewater than primary sludge. Unstabilized solids can be landfilled or incinerated. Stabilization can result in the production of biosolids.

Tertiary (Advanced) Wastewater Treatment

Used at POTWs that require higher effluent quality than that produced with secondary treatment. Common types of treatment include biological and chemical precipitation processes to remove nitrogen and phosphorus.	Lime, polymers, iron or aluminum salts used in tertiary wastewater treatment produce sewage sludge with varying water-absorbing characteristics. Also, high-level lime precipitation produces alkaline biosolids.

Source: Based on USEPA, 1999.

WASTEWATER TREATMENT AND BIOSOLIDS PRODUCTION

Wastewater treatment usually involves the following steps:

- Screening and grit removal
- Primary wastewater treatment
- Secondary wastewater treatment
- Tertiary (advanced) wastewater treatment
- Stabilization (biosolids production)
- Alkaline stabilization
- Anaerobic digestion
- Aerobic digestion
- Composting
- Heat drying and pelletizing

Table 1.8. A Summary of the Stabilization Processes that Produce Biosolids

Stabilization Process	USEPA 40CFR503 Classification and Use
Alkaline Stabilization	
Use of lime or other alkaline materials, such as cement kiln dust, lime kiln dust, Portland cement and fly ash. Increase of pH to reduce pathogens and achieve vector attraction reduction.	The process can produce a Class A or Class B. Major uses are in agriculture, reclamation, slope stabilization, structural fill, and municipal solid waste (MSW) landfills.
Anaerobic Digestion	
Anaerobic digestion involves biologically stabilizing biosolids in a closed vessel to reduce the organic content, mass, odor, and pathogens. During this process methane is generated and can be used as an energy source. Both mesophilic temperatures (35°C, 95°F) and thermophilic temperatures (55°C, 131°F) are used.	The process produces either a Class B at mesophilic temperatures or Class A at thermophilic temperatures. The production of Class B is more common. Principal uses are in agriculture and forestry.
Aerobic Digestion	
Aerobic digestion utilizes oxygen or air to biologically stabilize biosolids in an open or closed vessel or lagoon. The organic matter is converted to carbon dioxide, water, and nitrogen. Pathogens and odors are reduced. High thermophilic temperatures are recently being used.	Generally in most cases a Class B biosolids is produced. High temperature systems can produce a Class A product. The principal uses are in agriculture.
Composting	
Composting is the biological decomposition of the organic matter. Generally biosolids composting is done at thermophilic temperatures (>55°C, 131°F) in order to destroy pathogens. During composting the odorous compounds are reduced.	Composting produces a Class A product. Its principal uses are horticultural, including landscaping, nursery operations, turf, lawn and sod production, agriculture, public works department projects, highway beautification and reclamation.
Heat Drying and Pelletizing	
Heat drying involves using active or passive dryers to remove water from biosolids. It is used to destroy pathogens and remove water, which reduces the volume of material. In some cases heat-dried products are formed into pellets.	Heat drying produces a Class A biosolids product. It is primarily used in agriculture. Other uses are in turf and sod production and golf courses. Heat dried pellets are also blended with other nitrogen, phosphorus, and potassium chemicals to produce fertilizers.

An extensive discussion of the stabilization processes is provided by Switzenbaum et al., 1997. Other discussions can be found in several documents produced by USEPA (USEPA, 1994; 1999). Table 1.7 summarizes some key features of the stabilization processes that produce biosolids.

CONCLUSION

The advent of improved sanitation and wastewater treatment has resulted in the production of sewage sludge and biosolids. Increased public concern for environmental pollution of oceans, waterways, and the air has resulted in the increase in beneficial use. The increase in beneficial use, especially land application (including heat drying, composting, and alkaline stabilization), mandated that the quality of biosolids be improved. Industrial pretreatment requirements resulted in very significant reduction of pollutants. There were dramatic increases in research on heavy metals in the late 1970s, 1980s, and 1990s. This research allowed USEPA to promulgate risk-based regulations to protect public health and the environment.

The production of biosolids through various stabilization processes allows wastewater residuals to be beneficially used. The majority of the biosolids produced in the United States is land applied either directly to agriculture or further processed through composting or heat drying for use in agriculture or horticulture.

REFERENCES

Anderson, M.S., 1959, Fertilizing characteristics of sewage sludge, *Sewage Ind. Wastes,* 31: 678–682.

Bastian, R.K., 1997, The biosolids (sludge) management treatment beneficial use, and disposal situation in the USA, *Eur. Water Pollut. Control* 7(2): 62–78.

DeTurk, E.E., 1935, Adaptability of sewage sludge as a fertilizer, *Sewage Works J.* 7 (4): 597–610.

Doxiadis, C.A., 1973, Ancient Greek settlements: Third report, *Ekistics* 35(11).

Epstein, E. 1997, *The Science of Composting,* Technomic, Lancaster, PA.

Fuller, W.H., 1983, Soil injection of sewage sludge for crop production, Dept. Soils, Water and Eng., College of Agric., U. of Arizona, Tech. Bull. 250, Tucson.

Gerhard, W.P., 1909, *Sanitation and Sanitary Engineering,* self-published, New York.

Jewell, W.J. and B.L. Seabrook, 1979, A history of land application as a treatment alternative. U.S. Environmental Protection Agency, Technical Report EPA 430/9-79-012, Washington, D.C.

Matthews, P., 1999, Sewage sludge treatment and biosolids management in Europe, Conf. on Sewage Sludge Treatment and Disposal in Madrid.

Switzenbaum, M.S., L.H. Moss, E. Epstein, A.B. Pincince and J.F. Donovan. 1997. Defining biosolids stability, *J. Environ. Eng.* 123(12): 1178–1184.

USEPA, 1990, National Sewage Sludge Survey: Availability of information and data and anticipated impacts on proposed regulations; Proposed Rule, U.S. Environmental Protection Agency, *Fed. Reg.* 54(23): 47210–47283, Washington, D.C.

USEPA, 1993, Needs survey report to Congress, U.S. Environmental Protection Agency, Office of Water, EPA 832-R-93-002, Washington, D.C.

USEPA, 1994, A plain English guide to the EPA Part 503 Biosolids rule, Rep. No. EPA/832/R-93/003, U.S. Environmental Protection Agency, Office of Wastewater Management, Washington, D.C.

USEPA, 1999, Biosolids generation in the United States, U.S. Environmental Protection Agency, Municipal and Industrial Solid Waste Division, Office of Solid Waste, EPA530-R-99-009, Washington, D.C.

Webber, L.R. and B.C. Hillard, 1974, Agricultural use of sludge, 120-142. In *Proc. Conf. Program Abatement Munic. Pollut. Provis., Canada, Agreement Great Lakes Water Qual.,* University of Guelph, ON.

CHAPTER 2

Characteristics of Sewage Sludge and Biosolids

INTRODUCTION

The characteristics of biosolids play an important part in their use for land application. They can be broken down into three categories: physical, chemical, and biological. Physical properties affect the method of application, as well as the soil's physical and chemical properties. Several of these physical properties have an important effect on plant growth. They can affect the availability and accumulation of plant nutrients and trace elements. The important physical characteristics are:

- Solid content
- Organic matter content

Chemical properties affect plant growth as well as the soil's chemical and physical properties. The important chemical characteristics are:

- pH
- Soluble salts
- Plant nutrients — macro and micro
- Essential and non-essential trace elements to humans and animals
- Organic chemicals

Biological properties affect the soil's microbial population and organic matter's decomposition in soil. Biological characteristics also affect human health and the environment. All biosolids contain a wide variety of microbes. Many of these are very beneficial, while others can be harmful to humans, animals, or plants. The microbial population in biosolids is very important to the decomposition of organic matter. Since pathogens represent the most important microbial biosolids property for land application, this topic is covered in a separate chapter.

PHYSICAL PROPERTIES

The solids content of biosolids affects the method of land application. Liquid or low-solids biosolids will generally be injected into the soil to prevent vectors and provide better aesthetics. Vector reduction is part of the USEPA 40 CFR 503 regulations. The addition of liquid biosolids also increases the moisture content of the soil, which could benefit plant growth. The organic matter content is diluted and consequently its benefit in improving soil structure will occur only after repeated applications and after a long period of time. The amount of plant nutrients and trace elements depends on the quantity and percent solids of the biosolids.

Dewatered or semisolid biosolids are usually spread on the surface and subsequently plowed into the soil. The concentration of solids adds organic matter to the soil. This added organic matter improves the soil's physical properties, especially soil structure, soil moisture retention, soil moisture content, and cation exchange capacity. The dark content of the added organic matter could affect soil surface temperatures and in the spring hasten germination of crops. High solid biosolids are usually compost or heat-dried products. Compost is an excellent source of organic matter and will improve the soil's physical properties (Epstein, 1997), which include:

- Soil structure — bulk density, porosity, soil strength, and aeration (by increasing soil aggregation)
- Soil water relationships — water retention, available water to plants, and soil water content
- Water infiltration and permeability
- Soil erosion and runoff
- Soil temperature

Heat-dried products are generally applied as fertilizers and add little organic matter since small amounts are applied. They do not usually affect the soil's physical properties.

The organic content of biosolids will vary, depending on the solids content and extent of treatment. Organic matter in biosolids can be as high as 70% depending on the wastewater treatment (e.g., digestion, bulking agent addition in the case of compost, lime addition, and addition of other materials).

CHEMICAL PROPERTIES

The chemical properties of biosolids are affected by several factors:

- Quality of wastewater — extent of industrial pretreatment
- Extent of treatment — primary, secondary, tertiary
- Process modes — use of chemicals (e.g., ferric chloride, polymers, etc.)
- Methods of stabilization (e.g., lime treatment)

Table 2.1   Changes in Heavy Metal Concentration

Heavy Metal	Mean Concentration 40 Cities[1] mg/kg, dry wt.	Mean Concentration NSSS[2] Mg/kg, dry wt.	Part 503 Pollutant Concentration Limits mg/kg, dry wt.
Arsenic (As)	6.7	9.9	41
Cadmium (Cd)	69	6.94	39
Chromium (Cr)	429	119	1,200
Copper (Cu)	602	741	1,500
Lead (Pb)	369	134.4	300
Mercury (Hg)	2.8	5.2	17
Molybdenum (Mo)	18	9.2	–
Nickel (Ni)	135	42.7	420
Selenium (Se)	7.3	5.2	36
Zinc (Zn)	1,594	1,202	2,800

[1] USEPA, 1989.
[2] National Sewage Sludge Survey (USEPA, 1990).

Table 2.2   Trends in Metal Concentration of Sewage Sludge/Biosolids from 1976 to 1996

Year	As	Cd	Cr	Cu	Pb	Hg	Mo	Ni	Se	Zn
1976[1]	–	110	2620	1210	1360	–	–	320	–	2790
1979[2]	6.7	69	429	602	369	2.8	18	135	7.3	1594
1987[3]	12	26	430	711	308	3.3	19	167	6.0	1540
1988[4]	9.9	6.9	119	741	134	5.2	9.2	43	5.2	1202
1996[5]	12	6.4	103	506	111	2.1	15	57	5.7	830

[1] 150 treatment plants from cities in northeast and north central states (Sommers, 1977).
[2] Treatment plants from 40 cities (USEPA, 1989).
[3] Data for Cd, Cr, Cu, Pb, Ni, and Zn are from 62–64 U.S. treatment plants; As, Hg, Mo, and Se are from 37, 50, 12, and 30 plants, respectively (Pietz et al., 1998).
[4] 199 treatment plants from cities throughout the U.S. (USEPA, 1990).
[5] 203–210 treatment plants from cities throughout the U.S. (Pietz et al., 1998).
Source: Page and Chang, 1998.

Trace Elements, Heavy Metals, and Micronutrients

Biosolids contain trace elements, including heavy metals, primarily from industrial, commercial, and residential discharges into the wastewater system. As a result of the Clean Water Act of 1972 which restricted industrial discharge, the quality of the wastewater entering publicly owned wastewater treatment plants has improved significantly. In 1989, USEPA published data from 40 cities. Table 2.1 shows the changes in heavy metals as related to the 503 regulations. In the 40 cities study, cadmium (Cd) and lead (Pb) could not meet the Part 503 pollution limits. The effect of industrial pretreatment and regulations can be seen in Table 2.2. These data show the changes that have occurred from 1976 to 1996. Cadmium decreased from 110 mg/kg dry weight in 1976 to 6.4 mg/kg dry weight in 1996. The biggest change occurred after 1987. Similarly, large reductions occurred with the other heavy metals with the exception of copper (Cu). Copper did not change, probably since much of the copper entering wastewater treatment plants is from use of copper piping in the domestic system.

The addition of chemicals such as lime and ferric chloride can affect pH, composition, and chemical species. The chemical species could influence solubility and hence mobility in the soil or uptake by plants. The method of biosolids processing and stabilization also affects their characteristics (Richards et al., 1997). The authors evaluated the leachability of trace elements as indicated by the toxic characteristic leaching procedure (TCLP) of dewatered, composted, N-Viro, pellets, and incinerator ash. TCLP is commonly used to indicate potential leachability of metals. The data in Table 2.3 show that, with the exception of N-Viro, very small percentages of Cd, Cr, Cu, Mo, Ni, Pb, and Zn were extracted as a percentage of total content. This indicates that the potential for leaching and mobility in the soil is extremely low. State regulators that require the TCLP for biosolid products do not understand that this does not provide any information on mobility of heavy metals from composted and other products where the organic matter binds the heavy metals. It is applicable to salts. Also this procedure does not indicate uptake by plants.

Changes in materials used in domestic residences have also affected wastewater quality. Lead was used in early plumbing and is now prohibited. Copper piping has contributed Cu, especially when domestic water had a low pH. Considerable copper piping has reverted to plastic piping. Another source of heavy metals is from the food we eat and discharge of food materials into the wastewater stream. This is especially true where disposals are used.

Trace elements and heavy metals are ubiquitous. They are found in natural soils and plants. They are also in fertilizers since they are part of the mineralogical composition of the mined materials. This is especially true of many phosphate fertilizers that could contain high levels of cadmium and zinc. Raven and Loeppert (1997) analyzed 16 fertilizer materials. Three were primarily nitrogen (ammonium) products; eight were phosphate materials, and four were potassium sources. They also analyzed sewage sludge, organic materials, and liming materials. Among the fertilizers, phosphate sources had the highest heavy metals. Potassium and nitrogen fertilizers had insignificant amounts. Cadmium in phosphate fertilizers ranged from 0.7 to 48.8 µg/g; Cu from 0.68 to 19.6 µg/g; Ni from 0.6 to 50.4 µg/g; Pb from <0.2 to 29.2 µg/g, and Zn from not detected to 33.5. They concluded that trace and heavy metal concentrations generally decreased in this order: rock phosphate > biosolids > commercial phosphate fertilizer > organic amendments and liming material > commercial K fertilizers > commercial N fertilizers. As early as 1975, Lee and Keeny (1975) estimated that 2150 kg of Cd is added annually to Wisconsin soils through fertilizers and biosolids, with much more coming from fertilizers than biosolids.

Organic Compounds

Organic compounds are found in biosolids as a result of industrial and commercial discharges, household discharges, pesticides from runoff and soil. In 1980, USEPA reported on the occurrence and fate of 129 priority pollutants in the waste-

Table 2.3 Effect of Biosolids Stabilization on TCLP Extractability of Some Trace Elements

Mean Concentration - mg/kg

Trace Element	Dewatered Total	Dewatered Extract	Compost Total	Compost Extract	N-Viro Total	N-Viro Extract	Pellets Total	Pellets Extract	Ash Total	Ash Extract
Cd	5.62	ND	4.21	0.17	1.58	0.11	6.43	0.51	3.58	0.15
Cr	130	1.85	121	1.35	40	0.97	135	1.92	218	1.18
Cu	587	0.98	469	9.75	119	51.0	606	23.0	1219	21.5
Mo	49.7	0.56	32.7	0.55	9.8	4.93	55.3	1.50	95.1	4.52
Ni	35.8	2.54	32.5	1.58	12.7	3.07	38.0	5.84	74.8	3.66
Pb	132	0.81	109	0.10	NA	0.61	137	0.08	145	0.08
Zn	545	60.9	458	52.5	115	ND	567	84.7	959	39.1

NA = not available.

Source: Richards et al., 1997.

Table 2.4   Priority Pollutants Present in Combined Undigested Sewage Sludges at 20
            Wastewater Plants

| Organic Chemical | No. Times Detected | Concentration in Sludges | | | |
| | | µg/l, Wet | | mg/kg, Dry | |
		Median	Range	Median	Range
Bis (2-ethylhexyl) phthalate	13	3806	157–11257	109	4.1–273
Chloroethane	2	1259	517–2000	19	14.5–24
1,2-*trans*-Dichloroethylene	11	744	42–54993	21	0.72–865
Toluene	12	722	54–26857	15	1.4–705
Butylbenzyl phthalate	11	577	1–17725	15	0.52–210
2-Chloronaphthalene	1	400	400	4.7	4.7
Hexachlorobutadiene	2	338	10–675	4.3	0.52–8
Phenanthrene	12	278	34–1565	7.4	0.89–44
Carbon tetrachloride	1	270	270	4.2	4.2
Vinyl chloride	3	250	145–3292	5.7	3–110
Dibenzo(a,h)anthracene	1	250	25	13	13
Naphthalene	9	238	23–3100	7.5	0.9–70
Ethylbenzene	12	248	45–2100	5.5	1.0–51
Di-*n*-butylphthalate	12	184	10–1045	3.5	0.32–17
Phenol	11	123	27–4310	4.2	0.9–113
Methyl chloride	10	89	5–1055	2.5	0.06–30
Pyrene	12	125	10–734	2.5	0.33–18
Chrysene	9	85	15–750	2.0	0.25–13
Fluoroanthene	10	90	10–600	1.8	0.35--7.1
Benzene	11	16	2–401	0.32	0.053–11.3
Tetrachloroethylene	11	14	1–1601	0.38	0.024–42
Trichloroethylene	10	57	2–1927	0.98	0.048–44

Source: Naylor and Loehr, 1982, *BioCycle* 23(4): 18–22. With permission.

water and sludge from 20 publicly owned wastewater treatment plants in the United
States. Although the survey included small treatment plants, the median flow was
30.4 MGD, which represented a population of approximately 300,000 people (Naylor and Loehr, 1982). Table 2.4 shows the organic priority pollutants present in
combined undigested sewage sludges. Combined sludges consisted of a mixture of
sludges generated by two or more wastewater treatment processes (e.g., primary
plus secondary sludges). The authors felt that these data represented a conservative
(high) estimate of the actual amounts of organic priority pollutants, since no losses
could have occurred from digestion.

Jacobs et al. (1987) published an extensive list of organic chemicals found in
sewage sludges and biosolids. Table 2.5 is a summary of the data by chemical
group. As they indicated, sewage sludges and biosolids can be highly contaminated
with organic compounds. These data were cited prior to the implementation of
industrial pretreatment.

Subsequently, in 1990, USEPA published the results of the National Sewage Sludge
Survey that determined the chemical constituents in 209 wastewater treatment plants
randomly selected throughout the United States. The number of treatment plants in
which organic compounds were detected and the concentration of these compounds
were very low. Several of the pesticides, such as DDT and chlordane, have been banned

Table 2.5   Summary of Distribution of Organic Chemicals in Sewage Sludges and Biosolids by Chemical Groups

Chemical Group[1]	No. of Organic Chemicals Tested	No. of Organic Chemicals Tested Having Median Concentrations in Sludges and Biosolids mg/kg Dry Weight Basis				
		ND	<1	1-10	1-100	>100
Phthalate esters	6	0	0	1	4	1
Monocyclic aromatics	26	12	5	2	4	0
Polynuclear aromatics	7	0	4	2	1	0
Halogenated biphenyls	9	1	3	5	0	0
Halogenated aliphatics	10	0	6	4	0	0
Triaryl phosphate esters	3	0	0	2	1	0
Aromatic and alkyl amines	16	6	9	0	1	0
Phenols	12	0	1	11	0	0
Chlorinated pesticides and hydrocarbons	21	4	14	3	0	0
Miscellaneous	2	1	0	1	0	0
Totals	109	24	42	31	11	1

[1] There was inadequate data on dioxins and furans.

Source: Jacobs et al., 1987, pp. 101–143, A.L. Page et al., (Eds.), *Land Application of Sludge*, Lewis Publishers, Chelsea, MI. With permission.

from application and are no longer manufactured. Similarly PCBs are not being manufactured. However, both DDT and PCBs are very persistent in the environment.

The study gathered data at 180 publicly owned treatment works (POTWs), as well as survey data from 475 public treatment facilities with at least secondary wastewater treatment in the United States. USEPA screened 412 pollutants. These included dioxins/furans, pesticides, herbicides, semivolatile and volatile organic compounds. USEPA reviewed the scientific literature for toxicity, fate, effect, and transport information. The data showed extremely low levels; therefore, toxic organics were excluded from the 40 CFR 503 regulations. The data for the regulated priority pollutants are shown in Table 2.6 (USEPA, 1990). With the exception of Bis (2-ethylhexyl) phthalate, the other organics were essentially not detected. One reason for the low detection is the rather high detection limits.

Another group of organic chemicals of concern is surfactants, which are derived from detergent products, paints, pesticides, textiles, and personal care products (La Guardia et al., 2001). They are very abundant in biosolids, and concentrations range from 200 to 20,000 mg/kg dry weight (Haig, 1996). Three types of surfactant compounds (WEAO, 2001) are:

1. Anionic, e.g., linear alkylbenzene sulfonates (LAS), alkane ethoxy sulfonates (AES), secondary alkanesulfonates (SAS)
2. Nonionic, e.g., alcohol ethoxylates (AE), alkylphenols (AP), including alkylphenol polyethoxylates (APE)
3. Cationic, e.g., di-2-hydroxyethyl dimethyl ammonium chloride (DEEDMAC, quaternary esters)

Linear alkylbenzene sulfonates and alkylphenols are the most common surfactant compounds. Alkylphenols are endocrine disrupters. 4-Nonylphenols (NPs) are com-

Table 2.6   Organic Compounds Found in Biosolids

Organic Compound	Number of Times Detected	Mean (mg/kg)	Minimum (mg/kg)	Maximum (mg/kg)
Aldrin	8	0.029	0.019	0.046
Benzene	4	0.098	0.012	0.220
Benzo(a)pyrene	7	10.785	0.671	24.703
Bis(2-ethylhexyl)phthalate	189	107.233	0.510	89.129
Chlordane	1	0.489	0.489	0.489
4,4' –DDD	1	0.391	0.391	0.391
4,4' –DDE	4	0.100	0.030	0.190
4,4' –DDT	7	0.051	0.015	0.121
Dieldrin	6	0.024	0.013	0.047
Dimethyl nitrosamine	0	BDL[1]	BDL	BDL
Heptachlor	1	0.023	0.023	0.023
Hexachlorobenzene	0	BDL	BDL	BDL
Hexachlorobutadiene	0	BDL	BDL	BDL
Lindane (Gamma-BHC)	2	0.074	0.072	0.076
PCB-1016	0	BDL	BDL	BDL
PCB-1221	0	BDL	BDL	BDL
PCB-1232	0	BDL	BDL	BDL
PCB-1242	0	BDL	BDL	BDL
PCB-1248	23	0.740	0.043	5.203
PCB-1254	13	1.765	0.312	9.347
PCB-1060	20	0.671	0.031	4.006
Toxaphene	0	BDL	BDL	BDL
Trichloroethylene	7	0.848	0.024	3.302

[1] BDL = Below detection limit.
Source: USEPA, 1990

mon products of biodegradation of many nonionic surfactants, the nonylphenol ethoxylates (NPEs).

Guenther et al., (2002) analyzed 60 different food materials commonly available in Germany. They found that NPs were ubiquitous in foods. The concentrations of NPs ranged from 0.1 to 19.4 µg/kg wet weight basis, regardless of the fat content. They indicated that many of the sources in foods could be from packaging, cleaning agents, and pesticides. It is therefore not surprising that these compounds, in addition to entering the wastewater treatment plant from commercial and industrial sources, would also be deposited from food waste.

Some surfactants are biodegraded during biological treatment. Jensen (1999) reported that LAS compounds degrade very slowly or not at all under anaerobic conditions. Since more than 90% are removed from the liquid phase during waste-water treatment, significant amounts can be found in the solids portion. La Guardia et al. (2001) analyzed 11 biosolids and biosolid products, four Class A and seven Class B biosolids, for alkylphenol ethoxylate degradation products. These included octylphenol (OP), nonylphenols (NPs), nonylphenol monoethoxylates (NP1EOs) and nonylphenol diethoxylates (NP2EOs). As the authors indicate, these compounds are toxic and are endocrine disrupters.

Table 2.7 summarizes their data. In 10 of the 11 biosolids, nonylphenols were the most abundant of the byproducts. The mean concentration (722 mg/kg) in the

Table 2.7 Concentration of Alkyphenol Ethoxylate Degradation Products in Biosolids

Biosolid	Octylphenol	Nonylphenols	Nonylphenol Monoethoxylates	Nonylphenol Diethoxylates	Total
			mg/kg		
Class A					
Compost A	<0.5	5.4	0.7	<1.5	6.1
Compost B	1.5	172	2.5	<1.5	176
Compost C	<0.5	14.2	<0.5	<1.5	14.2
Heat dried	7.5	496	33.5	7.4	544
Class B					
Lime A	5.3	820	81.7	25.3	932
Lime B	2.0	119	154	254	529
Anaerobically digested	9.9	683	28.4	<1.5	721
Anaerobically digested	12.6	720	25.7	<1.5	758
Anaerobically digested	11.0	779	102	32.6	925
Anaerobically digested	11.7	707	55.8	<1.5	768
Anaerobically digested	6.7	8.7	64.9	22.7	981

Source: La Guardia et al., 2001.

anaerobically digested biosolids was nearly twice that of the heat-dried biosolids (496 mg/kg) and lime stabilized biosolids (470 mg/kg) and 12 times greater than the composted biosolids (64 mg/kg). The authors suggest that the lower values in the compost could be the result of dilution with bulking agents and further aerobic degradation. They report that degradation is greater under aerobic than anaerobic conditions.

Bennie (1999) reviewed the environmental occurrence of alkylphenols and alkylphenol ethoxylates in biosolids in Canadian wastewater treatment plants. He reported that the concentrations ranged from not detected (ND) to 850 mg/kg. The concentration of 4-NP ranged from 8.4 to 850 mg/kg; NP1EO from 3.9 to 437 mg/kg; NP2EO from 1.5 to 297; and NPnEO from 9 to 169. Octyl phenolics ranged from not detected to 20 mg/kg (Bennie, 1999; WEAO, 2001).

Another group of compounds found in biosolids that may be toxic to humans and animals is brominated diphenyl ethers (BDEs or PBDEs). Hale et al. (2001) examined 11 biosolid samples from California, New York, Virginia, and Maryland. The total concentrations ranged from 1,100 to 2,290 µg/kg dry weight basis. These are environmentally persistent compounds that have been found to bioaccumulate and be toxic in the aquatic environment. At the present time, the risk to humans is unknown.

One of the most toxic chemicals to animals and humans purportedly is a group of compounds termed dioxins. Dioxins are a group of congeners of chlorinated dibenzo-p-dioxins and dibenzofurans (Thomas and Spiro, 1996). Dioxins are ubiquitous and humans are exposed to them on a daily basis. In Round Two of the regulations, dioxins and some other organic compounds are being evaluated. This evaluation includes 29 specific congeners of polychlorinated dibenzo-p-dioxins, polychlorinated dibenzofurans, and coplaner polychlorinate biphenyls (PCBs). The agency is proposing a limit of 300 parts per trillion (ppt) toxic equivalents (TEQ) or nanograms TEQ per kilogram of dry biosolids.

Internationally, the median values of dioxin in sewage sludge generally ranges between 20 to 80 ng/kg TEQ (Carpenter, 2000). Jones and Sewart (1997) provided a comprehensive review of dioxins and furans in sewage sludges. They reported on the TEQ content of sewage sludges and biosolids from various countries. Their data are shown in Table 2.8.

Radionuclides may enter the sewage treatment plant principally as a result of discharges from medical facilities. They are relatively short lived (i.e., a short half-life). WEAO (2001) reported the medically used radionuclides most frequently observed were gallium-67, indium-111, iodine-123, iodine-131, thallium-201 and technetium-99.

Acidity (pH)

The pH of most biosolids — whether liquid, semisolid, or solid — is generally in the range of 7 to 8, unless lime is added during the wastewater treatment process. Lime, kiln dust and other alkaline products may be added to increase the pH and achieve the USEPA pathogen requirements. In some cases, such as in the biosolids

Table 2.8 Concentration of Dioxin in Sewage Sludges from Various Countries

Country	Number of Samples	Concentration – ng/kg Dry Weight		Source
		Range	Mean	
Germany	28	28–1560	102	Hagenmaier, 1988
Germany	13	20–177	47	Hagenmaier et al., 1992
Sweden	4	82–266	160	Broman et al., 1990
United States	239	0.49–2321	83	USEPA, 1990
England	11		150–200	DoE, 1989
England	16			DoE, 1993
Rural		9–73	23.3	
Mixed ind/rural		29–67	42.5	
Light ind/domestic		21–105	42.3	
Ind/domestic		7.6–192	52.8	
England	8	19–206	72	Sewart et al., 1995
Switzerland	30	6–4100	357	Rappe et al., 1994

Source: Jones and Sewart, 1997, Crit. Rev. Environ. Sci. Technol. 27(1): 1–85. With permission.

Table 2.9 Median Concentration of Several Macronutrients

Nutrient	Type of Biosolid			All Biosolids
	Anaerobic	Aerobic	Other	
% N	4.2	4.8	1.8	3.3
% P	3.0	2.7	1.0	2.3
% K	0.3	0.4	0.2	0.3
% Ca	4.9	3.0	3.4	3.9
% Mg	0.5	0.4	0.4	0.4
% Fe	1.2	1.0	0.1	1.1

Source: Sommers, 1977, J. Environ. Qual. 6: 225–232. With permission.

from the Washington, D.C. Blue Plains wastewater treatment plant, lime was added to reduce odors and avoid vectors during shipment to the composting facility at Site II in Maryland.

Plant Nutrients

Plant nutrients are among the most important chemical characteristics of biosolids. Farmers value biosolids for the nitrogen and phosphorus content. Because of this important characteristic, Chapter 3 is devoted to plant nutrients.

The major plant nutrients are nitrogen (N), phosphorus (P) and potassium (K). Other macronutrients are calcium (Ca), magnesium (Mg), and iron (Fe). Table 2.9 shows early data by Sommers (1977) for anaerobic, aerobic, and other biosolids. The other category included data from lagoons, primary, tertiary, and unspecified biosolids.

These early data do not reflect the use of alkaline products, which would increase the Ca and Mg content. Furthermore, during that period, ferric chloride and lime

Table 2.10 Plant Nutrients in Sewage Sludge and Biosolids from Several Cities

Plant Nutrient	Albuquerque 1987–1991	HRSD 1984–1991	MMSD 1989–1992	Denver 1983–1992	Chicago 1987–1990
% TKN	3.98–5.84	NA	7.4–10.6	4.9–9.5	7.83–8.85
% Organic N	NA	2.36–5.04	NA	NA	NA
% P	NA	1.72–3.0	2.5–2.8	2.12–3.4	8.32–10.89
% K	NA	0.14–0.28	0.6–1.0	0.30–0.50	1.08–2.24
% Ca	NA	NA	NA	NA	37.7–58.8
% Mg	NA	4.58–7.68	NA	NA	16.9–22.0
% Fe	9.29–17.7	NA			18.7–28.4

1987, 1988 data missing.

NA = Not available.

Source: Adapted from Stukenberg et al., 1993.

were used in dewatering, which was principally done by vacuum filters. Today belt filter presses and centrifuges predominate and use polymers. Consequently the concentration of Fe, Ca, and Mg would probably be lower. Later data for several biosolids are shown in Table 2.10.

BIOLOGICAL PROPERTIES

Microbiological

Pathogens, the most important biological property of biosolids for land application, are discussed in detail in Chapter 8. Another important characteristic is the indigenous microbial population. This population enhances the soil biota that is very important in organic matter decomposition and soil physical properties.

Many microbes and other organisms are involved in the decomposition process. There is very little data on this subject. This indigenous population consists of bacteria, fungi, actinomycetes, and protozoa. It has been reported that fresh animal manures may contain 10^6 anaerobic bacteria, 10^5 coliform bacteria, 10^6 enterococci bacteria, and 10^5 fungi per ml of suspension (Parr, 1974).

It is very possible that similar numbers are found in biosolids. Miller (1973) reported that application of biosolids increases the soil microbial population. Bacteria, actinomycetes, and fungi increased during the decomposition of anaerobically digested biosolids. Numbers of bacteria and actinomycetes were directly related to biosolids' loading rates. The increase in the fungal population was not as pronounced. A significant finding in this early study showed a change in the bacterial population from one dominated by Gram-positive bacteria in the unamended soil to one where Gram-negative bacteria were greater than 50%.

Organic Matter

Organic matter is an important constituent of biosolids. The use of biosolids for land application enhances the organic content of the soil. This is most important in sandy or clayey soils. In sandy soils the organic matter increases the water-holding capacity, soil aggregation, and other soil physical properties. It reduces the soil bulk density. Also, organic matter increases the cation exchange capacity, a very important property for supplying plant nutrients. The positive effect on the soil physical properties enhances the plant root environment. Plants are better able to withstand drought conditions, extract water, and utilize nutrients.

One of the earliest studies on the value of the organic matter content of biosolids was reported by Lunt (1953). He indicated that anaerobically digested biosolids provided modest increases (3% to 23%) in field moisture capacity, non-capillary porosity, and cation exchange capacity. The organic matter content increased by 35% to 40% and soil aggregation by 25% to 600%. The greatest increases occurred with a sandy soil, and less with loams.

Epstein (1973, 1975) indicated that organic matter through the activity of micro-organisms increases soil aggregation. Biosolid application increased the stable aggregates 16% to 33%. The addition of 5% sludge and biosolids increased the amount of water retained at different suction values. Adding biosolids and sludge increased the hydraulic conductivity of soils. Clapp et al. (1986) indicated that organic matter through the addition of biosolids reduced bulk density and increased total porosity and moisture retention of soils.

Lindsay and Logan (1998) evaluated the effect of anaerobically digested biosolids on soil physical properties when applied to a silt loam soil. They reported that bulk density significantly decreased, and porosity, moisture retention, percentage water stable aggregates, mean weight diameter of aggregates, liquid, and plastic limits increased with increasing biosolids application. Organic C increased linearly with biosolids application, and 4 years after application there was three times as much C in the high biosolids application rate. They concluded that the observed differences in soil physical properties are due to the effects of added organic matter and that these effects persisted for at least 4 years.

CONCLUSION

Pretreatment of industrial wastes discharged into domestic sewers has substantially reduced the levels of trace elements and heavy metals. USEPA regulations limiting heavy metal application to soils has also forced municipalities to enforce industrial and commercial discharges. Today, most biosolids have lower concentration of heavy metals than levels specified by USEPA in the 503 regulations. Domestic waters are major contributors of copper, especially in regions where the water entering homes is acidic.

Organic compounds are very low in biosolids. Furthermore, many organic compounds will be biodegraded in the soil. Organic compounds, because of their size, are generally not taken up by plant roots and translocated to the above-ground edible crop.

Biosolids enhance the microbial population, which increases the rate of organic matter decomposition in soils. As a result, there is a significant change in the soil physical properties. This produces a marked improvement in the plant root environment and better plant growth.

REFERENCES

Bennie, D.T., 1999, Review of the environmental occurrence of alkylphenols and alkylphenol ethoxylates, *Water Qual. Res. J. Can.* 34(1): 79–122.

Broman, D., C. Naf, C. Rolff, and Y. Zebuhr, 1990, Analysis of polychlorinated dibenzo-p-dioxins and polychlorinated dibenzofurans (PCDD/F) in soil and digested sewage sludges from Stockholm, Sweden, *Chemosphere* 21: 1213–1220.

Carpenter, A., 2000, Dioxin in organic residues, 14th Annual Residuals and Biosolids Management Conference, Boston, Water Environment Federation, Alexandria, Virginia.

Clapp, C.E., S.A. Stark, D.E. Clay, and W.E. Larson, 1986, Sewage sludge organic matter and soil properties, pp. 209–253. Y. Chen (Ed.), *The Role of Organic Matter in Modern Agriculture, Developments in Plant and Soil Sciences*, Martinus Nijhoff, Dordrecht, Netherlands.

DoE, 1989, Dioxins in the environment, Her Majesty's Stationary Office, Pollut. Paper No. 27, London, England.

DoE, 1993, The examination of sewage sludges for polychlorinated dibenzo-p-dioxins and polychlorinated dibenzofurans, Foundation for Water Res., FR/D 0009, London, England.

Epstein, E., 1973, The physical process in the soil as related to sewage sludge application, pp. 67–73. Recycling Municipal Sludges and Effluents on Land, Champaign, Illinois, National Assoc. State Universities and Land-Grant Colleges, Washington, D.C.

Epstein, E., 1975, Effect of sewage sludge on some soil physical properties, *J. Environ. Qual.* 4(1): 139–142.

Epstein, E. 1997, *The Science of Composting*, Technomic, Lancaster, Pennsylvania.

Guenther, K., V. Heinke, B. Thiele, E. Kleist, H. Prast, and T. Raecker, 2002, Endocrine disrupting nonylphenols are ubiquitous in food, *Environ. Sci. Technol.* 36(8): 1676–1680.

Hagenmaier, H., 1988, Investigation into PCDD/F levels and selected chlorohydrocarbons in sewage sludges, Federal Environmental Office, Germany.

Hagenmaier, H., J. She, T. Benz, N. Dawidowsky, L. Dusterhoft, and C. Lindig, 1992, Analysis of sewage sludge for PCDD/Fs and diphenylethers, *Chemosphere* 25: 1457–1462.

Haig, S.D., 1996, A review of the interaction of surfactants with organic contaminants in soil, *Sci. Total Environ.* 185: 161–170.

Hale, R.C., 2001, Persistent pollutants in land-applied sludges, *Nature* 412: 140–141.

Jacobs, L.W., G.A. O'Conner, M.A. Overcash, M.J. Zabik, and P. Rygiewicz, 1987, Effects of trace organics in sewage sludges on soil-plant systems and assessing their risk to humans, pp. 101–143, A.L. Page, T.J. Logan, and J.A. Ryan (Eds.), *Land Application of Sludge*, Lewis Publishers, Chelsea, Michigan.

Jensen, J., 1999, Fate and effects of linear alkylbenzene sulphonates (LAS) in the terrestrial environment, *Sci. Total Environ.* 226: 93–111.

Jones, K.C. and A.P. Sewart, 1997, Dioxins and furans in sewage sludges: A review of their occurrence and sources in sludge and of their environmental fate, behavior, and significance in sludge-amended agricultural systems, *Crit. Rev. Environ. Sci. Technol.* 27(1): 1–85.

La Guardia, M.J., R.C. Hale, E. Harvey and T.M. Mainor, 2001, Alkylphenol ethoxylate degradation products in land-applied sewage sludge (biosolids), *Environ. Sci. Technol.* 35(4): 4798–4804.

Lee, K.W. and D.R. Keeney, 1975, Cadmium and zinc additions to Wisconsin soils by commercial fertilizers and wastewater sludge application, *Water Air Soil Pollut.* 5: 109–112.

Lindsay, B. and T.J. Logan, 1998, Field response of soil physical properties to sewage sludge, *J. Environ. Qual.* 27: 534–542.

Lunt, H.A., 1953, The case for sludge as a soil improver, *Water Sewage* 100(8): 295–301.

Miller, R.H., 1973, Soil microbial aspects of recycling sewage sludges and waste effluents on land, pp. 79–90. Recycling Municipal Sludges and Effluents on Land, Champaign, Illinois, National Assoc. of State Universities and Land-Grant Colleges, Washington, D.C.

Naylor, L.M. and R.C. Loehr, 1982, Priority pollutants in municipal sewage sludge, *BioCycle* 23(4): 18–22.

Page, A.L. and A.C. Chang, 1998, Historical account of United States and other country standards for metals applied to land in the form of municipal sewage sludge, *Proceeding of the Society for Environmental Geochemistry and Health,* Hong Kong, China.

Parr, J.F., 1974, Organic matter decomposition and oxygen relationships, pp. 121–139. Factors Involved in Land Application of Agricultural and Municipal Wastes, Soil, Water and Air Sciences, Agricultural Res. Service, U.S. Department of Agriculture, Beltsville, Maryland.

Pietz, R.I., R. Johnson, R. Sustich, R. Granato, P. Tata, and C. Lue-Hing, 1998, A 1996 sludge survey of the Association of Metropolitan Sewerage Agencies Members, Metropolitan Water Reclamation District of Greater Chicago, Report No. 98-4, Chicago.

Rappe, C., R. Andersson, G. Karlaganis, and R. Bonjour, 1994, PCDDs and PCDFs in samples of sewage sludge from various source areas in Switzerland, *Organohalogen Compd.* 20: 79–84.

Raven, K.P. and R.H. Loeppert, 1997, Trace element composition of fertilizers and soil amendments, *J. Environ. Qual.* 26: 551–557.

Richards, B.K., J.H. Peverly, T.S. Steenhuis, and B.N. Liebowitz, 1997, Effect of processing mode on trace elements in dewatered sludge products, *J. Environ. Qual.* 26: 782–788.

Sewart, A., S.J. Harrad, M.S. McLachlan, S.P. McGrath, and K.C. Jones, 1995, PCDD/Fs and non-0-PCBs in digested U.K. sewage sludges, *Chemosphere* 30: 51–67.

Sommers, L.E., 1977, Chemical composition of sewage sludge and analysis of their potential use as fertilizers, *J. Environ. Qual.* 6: 225–232.

Stukenberg, J.R., S. Carr, L.W. Jacobs, and S. Bohm, 1993, Document long-term experience of biosolids land application programs, Water Environ. Res. Foundation, Project 91-ISP-4, Alexandria, Virginia.

Thomas, M.V. and T.G. Spiro, 1996, The U.S. inventory: Are there missing sources? *Environ. Sci. Technol.* 30(2): 82A–85A.

USEPA, 1989, Standards for the disposal of sewage sludge: proposed rule 40 CFR parts 257 & 503, Federal Register 54: 5746–5902, Washington, D.C.

USEPA, 1990, National sewage sludge survey: Availability of information and data, and anticipated impacts on proposed regulations, U.S. Environmental Protection Agency, Federal Register 55: 47210–47283, Washington, D.C.

WEAO, 2001, Fate and Significance of Selected Contaminants in Sewage Biosolids Applied to Agricultural Land through Literature Review and Consultation with Stakeholder Groups, Aurora, Ontario.

CHAPTER **3**

Plant Nutrients

INTRODUCTION

Plant nutrients can be subdivided into major nutrients and minor or micronutrients. Three of the major nutrients, nitrogen (N), phosphorus (P), and potassium (K), are the most important and are utilized to the greatest extent. These three plant nutrients are the ingredients listed on fertilizers sold at stores. Calcium (Ca), magnesium (Mg), and sulfur (S) are also major nutrients since the plant utilizes them to a greater extent than minor or micronutrients. The micronutrients essential to plant growth are boron (B), copper (Cu), iron (Fe), manganese (Mn), molybdenum (Mo), nickel (Ni), and zinc (Zn).

It has been recognized for centuries that sewage sludge and biosolids contain plant nutrients. Noer (1926) cited Bruttini (1923), who estimated that the annual excrement of the world's population contains approximately 9 million tons of nitrogen (N) and close to a million tons of phosphoric acid (P_2O_5) and potash (K_2O) that are lost to agriculture. Based on this data, Noer (1926) estimated that the annual loss to the United States is approximately 0.5 million tons of N valued at $180 million, and more than 100,000 tons each of P_2O_5, valued at $12 million, and K_2O, valued at $10 million.

Today this number would be much higher. It is estimated that the U.S. produces approximately 2 million dry tons of sludge and biosolids. This approximates to 10 million wet tons. Assuming an average of 4% nitrogen, the N in sewage sludge and biosolids would amount to about 400,000 tons that could be available to crops. Approximately 250,000 tons of P would be available to crops. In addition, significant quantities of minor elements Ca, Mg, and S would be available. In many cases, soils in the United States are deficient in these nutrients and farmers must add them at significant cost.

In 1977, Sommers obtained data on the nutrient content in 250 sewage sludge samples from 150 wastewater treatment plants. The data on the major plant nutrients, including calcium (Ca), magnesium (Mg), and sulfur (S), are presented in Table 3.1. Nitrogen, P, Ca, and S are present in relatively large amounts, whereas K and Mg are found in much smaller amounts.

Table 3.1 Concentrations of Total N, P, K, Ca, Mg and S in Sewage Sludge and Biosolids

Plant Nutrient	Sample				
	Type	Number	Range	Median	Mean
Total % N	Anaerobic	85	0.5–17.6	4.2	5.0
	Aerobic	38	0.5–7.6	4.8	4.9
	Other	68	<0.1–10.0	1.8	1.9
	All	191	<0.1–17.6	3.3	3.9
NH$_4$-N, ppm	Anaerobic	67	120–67,600	1,600	9,400
	Aerobic	33	30–11,300	400	950
	Other	3	5–12,500	80	4,200
	All	103	5–67,600	920	6,540
NO$_3$-N, ppm	Anaerobic	35	2–4,900	79	520
	Aerobic	8	7–830	180	300
	Other	–	–	–	–
	All	43	2–4,900	170	490
Total % P	Anaerobic	86	0.5–14.3	3.0	3.3
	Aerobic	38	1.1–5.5	2.7	2.9
	Other	65	<0.1–3.3	1.0	1.3
	All	189	<0.1–14.3	2.3	2.5
Total % K	Anaerobic	86	0.02–2.64	0.30	0.52
	Aerobic	37	0.08–1.10	0.38	0.46
	Other	69	0.02–0.87	0.17	0.20
	All	192	0.02–2.64	0.30	0.40
% Ca	Anaerobic	87	1.9–20.0	4.9	5.8
	Aerobic	37	0.6–13.5	3.0	3.3
	Other	69	0.1–25.0	3.4	4.6
	All	193	0.1–25.0	3.9	4.9
% Mg	Anaerobic	87	0.03–1.92	0.48	0.58
	Aerobic	37	0.03–1.10	0.41	0.52
	Other	65	0.03–1.97	0.43	0.50
	All	189	0.03–1.97	0.45	0.54
% S	Anaerobic	19	0.8–1.5	1.1	1.2
	Aerobic	9	1.6–1.1	0.8	0.8
	Other	–	–	–	–
	All	28	0.6–1.5	1.1	1.1

Source: Sommers, 1977, *J. Environ. Qual.* 6: 225–232. With permission.

The nitrogen cycle depicted in Figure 3.1 shows the various transformations that occur when biosolids are applied to land.

When biosolids are incorporated into the soil, the organic matter immediately begins to decompose by the microbes in the soil or those in the biosolids, as long as temperatures are above freezing and there is sufficient moisture. Even at low

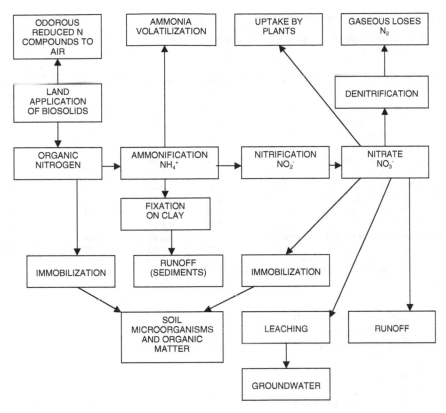

Figure 3.1   The nitrogen cycle when biosolids are applied to land. Many of these conditions
are interrelated. (From Epstein et al., 1978, *J. Environ. Qual.* 7: 217–221. With
permission.)

temperatures and under drought, decomposition proceeds at a very slow pace as a
result of the heat and moisture of the decomposing organisms. Microorganisms
utilize the carbon as a source of energy, and the nitrogen for cell development and
growth.

The nitrogen components in biosolids are predominantly organic. These have
been identified as proteinaceous, amino acids and hexosamines. Parker and Sommers
(1983) found that, for anaerobically digested biosolids, organic N ranged from 0.501
to 3.033%. Ammonium nitrogen (NH_4^+–N) ranged from 0.0026 to 0.3760%. Nitrate
nitrogen (NO_3^-–N) was extremely low. Their data are summarized in Table 3.2.

NITROGEN

When biosolids are applied to land, the predominantly organic nitrogen under-
goes numerous transformations. These transformations are extremely important as
they affect plant growth, microbial activity, and reactions through the soil. The
transformation of nitrogen in soil resulting from biosolids application is affected by
several important conditions in the biosolids and the soil. These include:

Table 3.2　Carbon and Nitrogen Characteristics of Biosolids

Plant No. and Type of Biosolids	Organic C	Inorganic C	Organic N	Inorganic N NH_4^+	Inorganic N $NO_2^- + NO_3^-$
	%	%	%	µg/g	µg/g
1. Aerobically digested	17.61	0.16	2.519	1360	44
1. Anaerobically digested	18.15	2.06	0.728	590	28
2. Aerobically digested	25.14	0.12	2.346	930	16
3. Aerobically digested	25.32	0.77	3.033	3760	33
4. Aerobically digested	17.47	1.67	1.654	340	1010
5. Anaerobically digested	27.58	1.05	2.744	2010	22
6. Anaerobically digested	11.75	1.87	1.048	56	780
7 Anaerobically digested	21.49	0.75	1.890	1490	120
8. Anaerobically digested	16.72	1.44	1.279	130	18
9. Anaerobically digested	6.78	5.18	0.501	26	90
10. Anaerobically digested	11.97	0.14	1.082	610	2100
11. Anaerobically digested	28.51	0.88	2.006	490	550
12. Anaerobically digested	16.08	0.16	1.692	2440	170
13. Anaerobically digested	15.32	1.66	1.403	630	32

Source: Adapted from Parker and Sommers, 1983.

- Moisture content of soil
- Temperature of soil
- Rate of mineralization
- Oxidation
- Aeration
- Soil porosity
- Biosolids characteristics
- Rate of microbial activity

The nature of the end products depends on the presence or absence of oxygen during the decomposition of the organic N in biosolids. In the presence of oxygen (aerobic conditions) the products are humus-like, nitrogen-containing products that become increasingly resistant to oxidation as decomposition progresses. Ammonia, which is released, can escape into the atmosphere or be converted to nitrites and then to nitrates. In the absence of oxygen (anaerobic conditions) the end products are humus-like materials, ammonia, and nitrogen gas.

Ammonification

Ammonification is the conversion of organic N into ammonium N (NH_4–N) by numerous heterotrophic soil organisms. This form of N is available to plants, but it generally does not leach through the soil, because the positively charged cation (NH_4^+) is held on the surface of negatively charged soil particles. This is termed fixation. Leaching can occur in sandy soils since the cation exchange capacity of these soils is low. NH_4–N attached to soil particles can contaminate surface waters through runoff and erosion.

Nitrification

Nitrification involves two steps. First NH_4^+ is oxidized to nitrite (NO_2^-) by *Nitrosomonas* spp. bacteria. Then NO_2^- is oxidized to nitrate (NO_3^+) by *Nitrobacter* spp. bacteria. Possibly other species are involved (Tate, 1995). Nitrate is readily available to plants. Nitrite-nitrogen is negatively charged and not adsorbed on soil particles. It thus remains in the soil solution and can be leached below the root zone and percolate into groundwater.

Immobilization

Immobilization is the tying up of nitrogen by soil organisms. When residues high in carbon are incorporated into the soil, the available N is used by the soil microorganisms that decompose the organic matter. Uncured biosolids compost, for example, when incorporated into the soil, will continue to decompose at a fairly high rate and the soil microorganisms will utilize the available N. This can sometimes be evident by yellowing (nitrogen deficiency) of plants. Although the N immobilized as microbial protein is later mineralized during decomposition, it can significantly reduce NO_3^- –N leaching during periods when crop uptake of N is reduced.

Biosolids-applied N may be immobilized into soil organic matter or remain as refractory organic N (Permi and Cornfield, 1969; Ryan et al., 1973).

Denitrification

Denitrification is the biological reduction of NO_3^- or NO_2 to gaseous forms of N, usually N_2O and N_2. The process is extremely rapid. The predominant bacteria are *Pseudomonas* or *Alcaligenes* (Tate, 1995). Under anaerobic conditions, the nitrogen in the soils is transformed into nitrogen gas and is released to the atmosphere. Since 80% of atmospheric gas consists of N, this addition of N is insignificant. This process primarily occurs in poorly aerated soils and waterlogged soils. After several days of waterlogged soils, much of the NO_3–N is lost by denitrification.

For denitrification to occur, decomposable organic matter must be present. Using liquid anaerobically digested biosolids can result in significant amounts of N losses by denitrification (Kelling et al., 1977). Several studies have indicated that denitrification is a major source of N loss when biosolids are applied (Kelling et al., 1977; Sommers et al., 1979).

Volatilization

When biosolids are applied to the soil surface, some volatilization of ammonia will occur. Any N volatilized is not available to plants or potentially available to be leached and contaminate groundwater. The amount of N volatilized from surface application of biosolids can vary significantly. Numerous studies attempted to evaluate the losses of N from land application of biosolids (Ryan and Keeney, 1975; Terry et al., 1978; Beauchamp, et al., 1978; Adamsen and Sabey, 1987; Robinson and Polglase, 2000). These studies indicate that the percentage of NH_4^+ –N can vary from less than 1% to 100%. The majority of N is volatilized during the first few days.

Robinson (1999) reported that when dewatered biosolids were applied in the field, 85% of NH_4^+ –N was lost within the first 3 weeks of application. In a subsequent study, Robinson and Polglase (2000) indicated that majority of NH_4^+ –N was lost during the first week and 71% to 81% was lost during the first 14 days.

Mineralization

Mineralization is the conversion of organic N to inorganic N. Organic N is first converted to NH_4^+ in a process termed ammonification. Then NH_4^+ is oxidized to NO_2^- by *Nitrosomonas* bacteria and NO_2^- is oxidized to NO_3^-. This process is called nitrification.

Thus, mineralization is the combination of ammonification and nitrification, whereby the organic N is converted to NH_4^+ –N and later to NO_3^- –N. The rate-limiting step in soil N mineralization is the conversion of organic N to NH_4^+ –N. Under conditions of adequate aeration and soil moisture and over a broad range of temperatures, the NH_4^+ –N is rapidly oxidized to NO_3^- –N (Stanford and Epstein, 1974).

Knowledge of the rate of nitrogen mineralization is important in determining the rate of biosolids application, potential for crop uptake, and potential for leaching. The amount of N mineralized is a function of several environmental factors and the type of biosolids applied. Chae and Tabatabai (1986) indicated that the rate of N mineralization is dependent on moisture, temperature, C:N ratio, and biosolids properties.

Figure 3.2 shows the rate of nitrogen mineralization for digested biosolids and composted biosolids as related to the amount of N applied per hectare (Epstein et al., 1978). Considerably more N was mineralized from the biosolids (41%) than from the compost (8.5%). The amount of N applied did not affect the percentage mineralized.

Table 3.3 summarizes the data obtained by numerous investigators. The variability shown in this table can be due to the type of biosolids and its condition, as well as the soils used in the studies. The extent of anaerobic or aerobic digestion and the initial N content will affect the organic N mineralization rate. Both composting and alkaline stabilization reduce the amount of N in biosolids, thus reducing the amount of N available for mineralization. Parker and Sommers (1983) indicate that immobilization of N may be a significant process in many soils treated with biosolids. They found that the amount of mineralizable N in biosolids was proportional to the total organic N. The anaerobic digestion or composting of primary, raw or waste activated sludges resulted in reduced organic N levels and decreased the potential amount of mineralizable N.

USEPA and the various states require that the rate of biosolids application be in relation to the crop requirement for nitrogen. This restriction is designed to avoid excess N and prevent leaching to groundwater. Consequently, the transformations that occur in the soil following biosolids application are extremely important. Nitrogen mineralization is the most important transformation since it releases inorganic

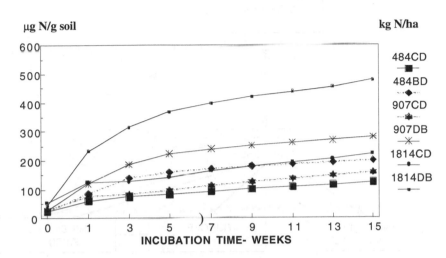

Figure 3.2   Inorganic nitrogen mineralization from digested biosolids (DB) and composted digested (CD) biosolids (From Epstein et al., 1978, *J. Environ. Qual.* 7: 217–221. With permission.)

Table 3.3   Nitrogen Mineralization Rates of Biosolids

Biosolids Type	Incubation Period	Percent Mineralized	Source
Anaerobically digested	16 weeks	4–48	Ryan et al., 1973
Anaerobically digested		36–41	Sabey et al., 1975
Anaerobically digested	13 weeks	14–25	Magdoff and Chromec, 1977
Anaerobically digested	15 weeks	40–42	Epstein et al., 1978
Aerobically digested	17 weeks	54–55	Magdoff and Amadon, 1980
Anaerobically digested	32 weeks	56.4–71.6	Lindmann and Cardenas, 1984
Composted digested	15 weeks	7–9.3	Epstein et al., 1978
Composted digested	54 days	6	Tester et al., 1977

N compounds which either are taken up by plants, immobilized, denitrified, or leached to groundwater. Therefore, it is important to know the potential mineralization rate for a given biosolids in relation to soil and climatic conditions.

PHOSPHORUS

Phosphorus (P) is an essential plant nutrient. It has been indicated that P deficiency is the second most important soil fertility problem throughout the world (Lindsey et al., 1989). However, excessive amounts of P in the soil tend to immobilize other chemical elements such as zinc (Zn) and copper (Cu) that are also essential

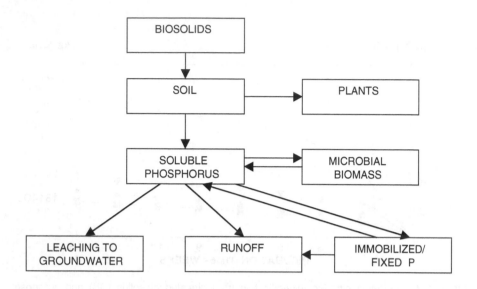

Figure 3.3   The phosphorus cycle when biosolids are applied to land.

for plant growth (Chang et al., 1983). Excessive P in soil can result in nonpoint source pollution of surface waters and shallow ground water (Sims et al., 2000).

Phosphorus in biosolids exists in both organic and inorganic forms. Organic P must undergo mineralization in the soil before plants can take it up. Figure 3.3 shows the P cycle when biosolids are applied to land. Inorganic P is predominant in biosolids. When biosolids are applied at rates consistent with the nitrogen requirement of the crop, excessive P often is applied. Pierzynski (1994) calculated that a biosolid having 13 g/kg plant available N (PAN) and 10 g/kg total P applied to supply 150 kg PAN/ha would also apply 115 kg P/ha. This amounts to approximately three times more than would be typically recommended for corn. Maguire et al. (2000) indicated that adding biosolids at currently recommended rates based on PAN will lead to an accumulation of P in soils. Sui et al. (1999) also concluded that adding biosolids at an application rate based on potentially available N could result in the accumulation of P and potentially be of environmental concern. This accumulation of P can result in eutrophication and potentially impact water bodies through surface runoff, subsurface drainage water, or eroded soil. They applied biosolids at rates of 0, 7.4 and 13.0 Mg dry matter per hectare. After 6 years of continuous application to poplars, total P increased significantly at both the 0 to 5 and 5 to 25 cm depths.

POTASSIUM

Potassium in biosolids is generally low (Table 3.1). Since K compounds in the wastewater are soluble, they generally do not settle in biosolids. When biosolids are land applied, the exchangeable form of K is the primary source for plants. This element is essential for plant growth and is in sufficient short supply in the soil to limit crop yield. In plants, it is important in amino acid and protein synthesis and photosynthesis. Excess of K in soil can reduce the uptake of other important cations by plants.

MICRONUTRIENTS

Micronutrients are elements essential to plant growth that are found in trace amounts in soils. Many of these micronutrients, however, can be toxic to humans, animals and plants. Several of them are also among the regulated heavy metals because they can be toxic to humans and animals. The seven micronutrients in biosolids are boron (B), copper (Cu), Iron (Fe), manganese (Mn), molybdenum (Mo), nickel (Ni), and zinc (Zn). The importance of micronutrients in plant growth and agriculture has been extensively studied. The role of these micronutrients, which are regulated as heavy metals, is discussed in Chapter 4. A major resource for further information is *Micronutrients in Agriculture* (Mortvedt et al., 1991).

Early data by Sommers (1977) show the concentration of micronutrients in sewage sludge and biosolids (see Table 3.4).

Boron has several functions in plants. It is important in cell metabolism, nucleic acid, and plasma membrane function (Römheld and Marschner, 1991). The concentration range between B deficiency and toxicity is very small. Boron may be essential to animals and humans, but its nutritional importance has not been established.

Copper is a regulated heavy metal that is essential to plants, animals, and humans. Its toxicity and essentiality are discussed in Chapter 4. In plants, it is important in many roles including photosynthesis, respiration, enzymes, nodulation in legumes, flowering and senescence, reproductive growth, and seed and fruit yield (Römheld and Marschner, 1991).

Table 3.4   Concentration of Plant Micronutrients in Sewage Sludge and Biosolids

Micronutrient	Number of Samples	Range	Median mg/kg	Mean
Boron	109	4–760	33	77
Copper	205	84–10,400	850	1210
Manganese	143	18–7100	260	380
Molybdenum	29	5–39	30	28
Nickel	165	2–3,520	82	320
Zinc	208	101–27,800	1,740	2790
		Percent		
Iron	165	<0.1–15.3	1.1	1.3

Source: Sommers, 1977, *J. Environ. Qual.* 6: 225–232. With permission.

Iron is essential to plants, animals, and humans. In plants, it is a component of proteins. It affects photosynthesis, respiration, sulfur reduction, and nitrogen fixation (Römheld and Marschner, 1991).

Manganese is essential to plants and is involved in photosynthesis. It also is important in enzyme functions and reactions. It appears to play key roles in disease resistance and plants' defense mechanisms (Römheld and Marschner, 1991).

Molybdenum is important in nitrogen metabolism of plants. It is contained in enzymes such as nitrate reductase and sulfate reductase as well as nitrogenase, and xanthine dehydrogense. It affects pollen formation, tasseling and development of anthers in corn, flowering delay, and fruit production (Römheld and Marschner, 1991).

Nickel's essentiality to plants is comparatively recent. It is important for certain enzymes such as urease. Brown et al. (1987) showed that Ni was essential for grain viability in barley. Because its essentiality to plants has only recently been discovered, there is little research on its importance in plant growth.

Also essential to plants, zinc is involved in carbohydrate metabolism as well as proteins and auxins. This element is important in stabilization and structural orientation of certain membrane proteins (Römheld and Marschner, 1991). Zinc deficiency has been extensively studied. Many soil areas in the United States are zinc deficient and supplemental Zn is often added to soils.

CONCLUSION

Biosolids can be an important source of plant macro and micronutrients for agricultural or horticultural crops, forestry, and land reclamation. The macronutrients nitrogen and phosphorus can provide the required amounts needed by crops. Potassium, the third most important macronutrient, is low in biosolids, which can also be a source of iron, boron, copper, nickel and other micronutrients essential for plant growth.

Excessive nitrogen or phosphorus applied to soils can be a source of pollution in both ground- and surface water. Nitrogen compounds can move through the soil profile into groundwater resources, and phosphorus and nitrogen in runoff and eroded particles can enter surface waters, resulting in eutrophication.

REFERENCES

Adamsen, F.J. and B.R. Sabey, 1987, Ammonia volatilization from liquid digested sewage sludge as affected by placement in soil, *Soil Sci. Soc. Am. J.* 51: 1080–1082.

Beauchamp, E.G., G.E. Kiddand and G. Thurtell, 1978, Ammonia volatilization from sewage sludge applied in the field, I. *Environ. Qual.* 7: 141–146.

Brown, P.H., R.M. Welch and E.E. Cary, 1987, Nickel: A micronutrient essential for higher plants, *J. Plant Nutr.* 10: 2125–2135.

Bruttini, A., 1923, *Uses of Waste Materials*, P.S. King and Son, Ltd., London, U.K.

Chae, Y.M. and M.A. Tabatabai, 1986, Mineralization of nitrogen in soils amended with organic wastes, *J. Environ. Qual.* 15: 193–198.

Chang, A.C., A.L. Page, F.H. Sutherland and E. Grgurevic, 1983, Fractionation of phosphorus in sludge-affected soils, *J. Environ. Qual.* 12(2): 286–290.

Epstein, E., D.B. Keane, J.J. Meisinger and J.O. Legg, 1978, Mineralization of nitrogen from sewage sludge and sludge compost, *J. Environ. Qual.* 7: 217–221.

Kelling, K.A., L.M. Walsh, D.R. Keeney, J.A. Ryan and A.E. Peterson, 1977, A field study of the agricultural use of sewage sludge: II. Effect on soil N and P, *J. Environ. Qual.* 6(4): 345–351.

Lindmann, W.C. and M. Cardenas, 1984, Nitrogen mineralization potential and nitrogen transformations of sludge-amended soil, *Soil Sci. Soc. Am. J.* 48: 1072–1077.

Lindsay, W.L., P.L.G. Vlek and S.H. Chien, 1989, Phosphate minerals, 1089–1130. *Minerals in Soil Environments*, 2nd ed., Soil Science Society of America, Madison, WI.

Magdoff, F.R. and F.F. Amadon, 1980, Nitrogen availability from sewage sludge, *J. Environ. Qual.* 9: 451–455.

Magdoff, F.R.and Chromec, 1977, Nitrogen mineralization from sewage sludge, *J. Environ. Sci. Health*, A12: 191–201.

Maguire, R.O., J.T. Sims and F.J. Coale, 2000, Phosphorus solubility in biosolids-amended farm soils in the Mid-Atlantic region of the USA, *J. Environ. Qual.* 29: 1225–1233.

Mortvedt, J.J., F.R. Cox, L.M. Shuman and R.M. Welch, 1991, *Micronutrients in Agriculture*, 2nd ed., Soil Science Society of America, Madison, WI.

Noer, O.J., 1926, Activated sludge: Its production, composition and value as fertilizer, *J. Am. Soc. Agron.*, 18 (11): 953–962.

Parker, C.F. and L.E. Sommers, 1983, Mineralization of nitrogen in sewage sludge, *J. Environ. Qual.* 12 (1): 150–156.

Pierzynski, G.M., 1994, Plant nutrient aspects of sewage sludge, 21–25, in C.E. Clapp, W.E. Larson and R.H. Dowdy (Eds.), *Sewage Sludge: Land Utilization and the Environment*, American Society of Agronomy, Crop Science Society of America and Soil Science Society of America, Madison, WI.

Premi, P.R. and A.H. Cornfield, 1969, Incubation study of nitrification of digested sludge added to soil, *Soil Biol. Biochem.* 1: 1–4.

Robinson, M.B., 1999, Nitrogen Dynamics after Application of Biosolids to a *Pinus radiata* Plantation, Ph.D. thesis, University of Melbourne, Australia.

Robinson, M.B. and P.J. Polglase, 2000, Volatilization of nitrogen from dewatered biosolids, *J. Environ. Qual.* 29: 1351–1355.

Römheld, V. and Marschner, H., 1991, Function of micronutrients in plants, 297–328, in J.J. Mortvedt, F.R. Cox, L.M. Shuman and R.M. Welch (Eds.), *Micronutrients in Agriculture*, Soil Science Society of America, Madison, WI.

Ryan, J.A. and D.R. Keeney, 1975, Ammonia volatilization from surface applied wastewater sludge, *J. Water Pollut. Control Fed.* 47: 386–393.

Ryan, J.A., D.R. Keeney and L.M. Walsh, 1973, Nitrogen transformations and availability of anaerobically digested sewage sludge in soil, *J. Environ. Qual.* 2: 489–492.

Sabey, B.R., N.N. Agbim and D.C. Markstrom, 1975, Land application of sewage sludge: III. Nitrate accumulation and wheat growth resulting from addition of sewage sludge and wood wastes to soils, *J. Environ. Qual.* 4: 388–393.

Sims, J.T., A.C. Edwards, O.F. Schoumans and R.R. Simard, 2000, Integrating soil phosphorus testing into environmentally based agricultural management practices, *J. Environ. Qual.* 29: 60–71.

Sommers, L.E., 1977, Chemical composition of sewage sludges and analysis of their potential use as fertilizers, *J. Environ. Qual.* 6 (2): 225–232.

Sommers, L.E., D.W. Nelson and D.J. Silviera, 1979, Transformations of carbon, nitrogen and metals in soils treated with waste materials, *J. Environ. Qual.* 8(3): 287.

Stanford, G. and E. Epstein, 1974, Nitrogen mineralization–water relations in soil, *Soil Sci. Soc. Am. Proc.* 38: 103–107.

Sui, Y., M.L. Thompson and C.W. Mize, 1999, Redistribution of biosolids-derived total phosphorus applied to a Mollisol, *J. Environ. Qual.* 28: 1068–1074.

Tate, R.L. III, 1995, *Soil Microbiology*, John Wiley & Sons, New York.

Terry, R.E., D.W. Nelson, L.E. Sommers and G.J. Meyer, 1978, Ammonia volatilization from wastewater sludge applied to soils, *J. Water Pollut. Control. Fed.* 50: 2657–2665.

Tester, C.F., L.J. Sikora, J.M. Taylor and J.F. Parr, 1977, Decomposition of sewage sludge compost in soil: I. Carbon and nitrogen transformation, *J. Environ. Qual.* 6(4): 459–462.

CHAPTER 4

Trace Elements:
Heavy Metals and Micronutrients

INTRODUCTION

Trace elements are required in small amounts by plants or animals. Some of these have been identified while others may still be unknown. Heavy metals are a group of elements found in the periodic table with a relatively high molecular weight (density >5.0 mg/m^3) and, when taken into the body, can accumulate in specific body organs. Ashworth (1991) argues that the term "heavy metals" is a misnomer, because at least two elements, arsenic and selenium, are not metals. The trace elements often referred to as heavy metals that have been regulated are: arsenic (As), cadmium (Cd), copper (Cu), lead (Pb), mercury (Hg), molybdenum (Mo), nickel (Ni), selenium (Se) and zinc (Zn). Chromium (Cr) was regulated in the first draft of the 503 regulations issued in 1993. In 1995, Cr was deleted.

In this chapter, the term *trace elements* will be used except where the literature specifically uses the term *heavy metals*.

Micronutrients are those essential trace elements that are needed in relatively small quantities for growth of plants, animals, or humans. The eight plant micronutrients are: boron (B), copper (Cu), iron (Fe), manganese (Mn), molybdenum (Mo), nickel (Ni), selenium (Se) and zinc (Zn) (Mortvedt et al., 1991). Elements such as nitrogen (N), phosphorus (P), potassium (K), calcium (Ca) and magnesium (Mg) are referred to as macronutrients because they are required in large amounts by plants. The elements cobalt (Co), iodine (I), Cu, Fe, Mn, Mo, Se and Zn are trace elements essential for animal nutrition (Miller et al., 1991).

Several other elements — arsenic, boron, bromine, cadmium, lithium, nickel, lead, silicon, tin and vanadium — have more recently been proposed as essential to some animal species (Van Campen, 1991). Van Campen identifies eight essential trace elements for human nutrition: Cu, Cr, Fe, I, Mn, Mo, Se and Zn.

The heavy metals indicated in the USEPA and state regulations are trace elements that can be harmful to the environment, humans, animals and plants. Consequently, both regulations and literature rarely consider whether these elements are also essential to humans, animals, or plants. At many agricultural areas in the United

Table 4.1   Trace Metal Content of Yard Waste

Heavy Metal	Number of Samples	Mean mg/kg	SD	Min. mg/kg	Max. mg/kg	CV
Arsenic	5	4.8	5.05	1	12.8	106.16
Boron	30	28.7	17.93	0.2	76	62.41
Cadmium	29	0.32	0.20	0.04	0.81	62.22
Chromium	35	39.4	45.41	3.7	236	115.39
Copper	35	64	65.47	8	327	102.15
Lead	35	69.6	54.49	11.4	235	78.34
Mercury	22	0.19	0.11	0.04	0.5	59.57
Molybdenum	17	0.22	0.32	0.05	1.09	143.59
Nickel	33	26.89	28.27	3.27	152	105.13
Selenium	17	0.33	0.10	0.1	0.55	31.89
Zinc	35	153.0	74.13	41.6	295	48.47

Source: Epstein, 1997, The Science of Composting, Technomic, Lancaster, PA. With permission.

States, farmers apply small quantities of a trace element, which is also regulated as a heavy metal. Horticulturalists often add trace elements needed for plant nutrition even though several are considered heavy metals by regulators.

Biosolids contain trace elements as a result of atmospheric deposition on land, natural vegetation, food sources (because plant material will contain trace elements), industrial sources, fertilizers and pesticides, human wastes (due to ingestion of food and water) and natural soil. All of these materials can find their way into the sewer system and eventually end in the wastewater treatment plant and into biosolids. As an example, Table 4.1 shows the concentration of trace elements in yard waste.

The subject of trace elements in biosolids and their impact on human health and the environment has been very extensively studied over the past 30 years. Two chapters are devoted to this subject. This chapter covers environmental aspects and human health, while Chapter 5 discusses soil–plant interactions.

The objectives of these chapters are to:

• Provide data on the sources of trace elements, heavy metals and micronutrients in the environment
• Discuss the toxicology of trace elements
• Discuss the fate of trace elements in soils as they relate to plant uptake and the environment
• Provide information on uptake of trace elements by plants.

SOURCES OF TRACE ELEMENTS, HEAVY METALS, AND MICRONUTRIENTS IN THE ENVIRONMENT

Soils are derived from parent material as a result of weathering. Because many of the parent material minerals contain trace elements, natural soils will contain different amounts of trace elements depending on the type of mineral. Krauskopf (1967) reported that shale contained 6.6 mg/kg As; 0.3 mg/kg Cd; 57 mg/kg Cu; 20

Table 4.2   Trace Elements in Natural Soils, Agricultural Soils and Fertilizers in the
United States

Element	Range in Natural Soil[1] mg/kg	Range in Agricultural Soils[2] mg/kg	Range in Fertilizers[3] mg/kg
Arsenic	5–13	NA	0.3–1662.3
Cadmium	0.01–7	<0.0010–2.0	0.75–398
Chromium	23–15,000	NA	1.3–338.9
Copper	1–300	<0.6–495	1.0–29,650
Lead	2.6–25	<1.0–135	4.6–10,013
Mercury	NA	NA	0.011–3.36
Nickel	3–300	0.7–269	1.4–890
Selenium	0.0001–3.4	NA	NA
Zinc	10-2,000	<3.0–264	1.6–77,300

[1] Based on Conner and Shacklett, 1975
[2] Holmgren et al., 1993
[3] Moss et al., 2002
NA – not available

mg/kg Pb; and 80 mg/kg Zn. The values for granite were 1.5 mg/kg As; 0.2 mg/kg Cd; 10 mg/kg Cu; 20 mg/kg Pb; and 40 mg/kg Zn.

In Minnesota, soils developed from lacustrine clays (formed in lakes) have a higher level of Cd than other soils (Pierce et al., 1982). Arsenic occurs in more than 200 naturally occurring minerals (Onken, 1995). One of the major agricultural production areas in California, Salinas Valley, contains high levels of Cd due to a natural geological source: the Monterey shale. Cadmium concentrations in the surface soils ranged from 1.4 to 22 µg/g with an average 8.0 µg/g (Lund et al., 1981).

Many agricultural soils may have higher levels of heavy metals than normally found in natural soils as the result of atmospheric deposition and application of fertilizers, pesticides and biosolids. Haygarth et al. (1995) reported that from 30% to 53% of Se found on pasture leaves resulted from atmospheric deposition. Several other researchers have reported on significant deposition of Pb, Cd, As, Cu and Zn (Haygarth et al., 1995; Hovmand et al., 1983; Berthelsen et al., 1995; Harrison and Chirgawi, 1989).

Mortveldt et al. (1981) reported on the uptake of Cd by wheat from phosphorus fertilizers. Lee and Keeney (1975) found that the application of fertilizers added more Cd and Zn to soils in Wisconsin than biosolids at that time. Table 4.2 shows the heavy metal content of natural soils, agricultural soils and fertilizers (Conner and Shacklett, 1975; Holmgren et al., 1993). Mermut et al. (1996) reported that phosphate fertilizers can be a significant source of trace elements and suggested that some of these elements, especially Cd, Cr and Zn, can be a source of soil pollution.

In 1997, Washington State published a survey on heavy metals in fertilizers and industrial by-product fertilizers (Bowhay, 1997). Table 4.3 summarizes some of the data. Although the level of many heavy metals and other trace elements can be low in agricultural fertilizers, repeated applications over long periods of time could result in significant uptake and accumulation by food crops.

Table 4.3   Concentration of Heavy Metals in Some Fertilizers and Industrial By-Product Fertilizers

Element	Number of Samples Detected	Range in Concentration mg/kg
Arsenic	36	4.2–1,040
Cadmium	12	0.63–275
Copper	36	0.094–39,900
Lead	11	2.5–11,300
Mercury	36	0.006–3.36
Molybdenum	14	1.3–17.8
Nickel	16	1.5–195
Selenium	36	Not detected
Zinc	36	0.21–203,000

Source: Adapted from Bowhay, 1997.

Arsenic has been used as a defoliant for several crops prior to the 1980s and is still used in cotton. Blueberry and potato soils in Maine, where arsenic has been used as a defoliant, showed an increase in the level of this element. Lead arsenate and calcium arsenate previously have been used in cotton and orchards (Woolson et al., 1971). Many urban soils contain high levels of Pb as a result of lead-based gasoline or paints. Because Pb does not move readily through the soil, high levels will remain in surface soils for many years. Holmgren et al. (1993) analyzed 3,045 surface soil samples throughout the United States. Table 4.4 shows a summary of the data. Holmgren et al. found regional as well as local differences due to soil parameters. Soil Cd was lower in the southeast and generally higher in California, Michigan and New York. Organic soils had higher amounts that might have been the result of heavy application of phosphate fertilizers used in intensive vegetable production.

Low levels of Pb were found in the southeast. Some areas in Virginia and West Virginia had levels exceeding 3000 mg/kg. High levels of Pb were also found in the Ohio, Mississippi and Missouri River valleys. Some have suggested that the high levels may have been a result of industrial contamination.

Zinc levels were low in the southeast with moderately high levels in California, the southwest, Colorado and the lower Mississippi valley. Copper levels were also lower in the southeast with the exception of Florida. High levels were found in organic soils used for vegetable production in Florida, Michigan and New York, presumably as a result of fertilizer applications to correct Cu deficiency. Ma et al. (1997) reported much lower metal contents in 40 mineral soils of Florida. Organic soils had considerably higher levels of heavy metals than mineral soils. The higher the clay content, the higher the metal concentration.

Dudas and Pawluk (1980) determined the background levels of As, Cd, Co, Cu, Pb and Zn in Chernozemic and Luvisolic soils from Alberta, Canada. Arsenic ranged

Table 4.4   Geometric Means for Some Heavy Metals in U.S. Agricultural Surface Soils by Soil Texture

Soil Texture	Number of Samples	Cd	Zn	Cu	Ni	Pb
			mg/kg Dry Soil			
Loamy sand	384	0.055 g*	14.9 f	6.0 h	6.2 h	5.5 h
Sandy loam	208	0.096 f	26.1 e	10.8 g	11.6 fg	8.3 f
Fine sandy loam	308	0.107 f	28.3 e	10.3 g	12.1 fg	7.3 g
Silt	745	0.185 e	50.4 d	18.1 f	12.4 d	13.4 g
Loam	326	0.199 e	48.4 d	18.6 f	20.6 e	10.6 e
Silty clay loam	322	0.288 d	76.9 b	28.7 d	35.5 c	16.0 c
Clay	108	0.289 d	98.0 a	37.6 c	52.0 a	17.7 ab
Clay loam	148	0.294 d	65.3 c	22.7 e	28.4 d	12.1 d
Silty clay	59	0.388 c	97.7 a	33.6 c	43.1 b	16.4 bc
Muck	190	0.558 b	65.3 c	75.8 b	11.2 g	10.9 c
SAPRIC	88	0.811 a	59.7 c	97.9 a	12.8 f	18.3 a
All	2886	0.178	43.2	18.3	16.9	10.5

* Means within a column followed by the same letter are not statistically significant.
Source: Holmgren et al., 1993, J. Environ. Qual. 22: 335–348. With permission.

from 0.82 to 6.9 mg/kg; Cd from 0.53 to 0.6 mg/kg; Co from 6.4 to 15 mg/kg; Cu from 11 to 49 mg/kg; Pb from 15 to 41 mg/kg; and Zn from 29 to 235 mg/kg. Cd and Zn levels in Canadian soils were higher than those reported by Holmgren et al. (1993) for U.S. agricultural soils.

It is very evident from these data that trace elements, including heavy metals, are found universally in our environment.

TRACE ELEMENTS IN BIOSOLIDS

Biosolids contain trace elements and heavy metals primarily from industrial, commercial and residential discharges into the wastewater system. As a result of the Clean Water Act of 1972 restricting industrial discharge, the quality of the wastewater entering publicly owned treatment works (POTW) systems has improved. Consequently, the quality of biosolids has improved. Changes in materials used in domestic residences have also affected wastewater quality. Lead was used in early plumbing and is now prohibited. To a large extent, plastic piping has replaced copper piping.

Table 4.5 compares the heavy metals from an early 40-city POTW study conducted in 1979-80 to the 1988-89 National Sewage Sludge Survey (NSSS). Technically the data are not comparable. However Cd, Cr, Pb and Ni were greatly reduced. There was little change in Zn and Cu (USEPA, 1990). A comparison between U.S. and Canadian heavy metal concentrations is shown in Table 4.6 (based on a report prepared for the Water Environment Association of Ontario, 2001).

Industrial pretreatment in many of the large cities resulted in major reductions in heavy metals.

Table 4.5   A Comparison of Heavy Metal Concentrations in 40 POTWs in 1980
            to the NSSS Study in 1988

Element	Samples	Percent Detected	Mean mg/kg	SD	Coefficient of Variation
Arsenic	199	80	9.9	18.8	1.9
	45	100	6.7	6.59	0.98
Cadmium	198	69	6.94	11.8	1.69
	45	100	69.0	252	3.65
Chromium	199	91	119	339.2	2.86
	45	100	429	440.8	1.03
Copper	199	100	741	961.8	1.30
	45	100	602	528.8	0.88
Lead	199	80	134.4	197.8	1.47
	45	100	369	331.5	0.90
Mercury	199	63	5.2	15.5	2.98
	45	100	2.8	2.6	0.93
Molybdenum	199	53	9.2	16.6	1.79
	45	75	17.7	16.7	0.94
Nickel	199	66	42.7	94.8	2.22
	45	100	135.1	169.1	1.25
Selenium	199	65	5.2	7.3	1.42
	45	100	7.3	29.10	4.16
Zinc	199	100	1,202	1,554.4	1.29
	45	100	1,594	1,759.3	1.10

Source: USEPA, 1990.

Table 4.6   Comparison of Heavy Metal Concentration in
            United States and Canadian Biosolids

Element	United States Surveys mg/kg Dry Weight			Canadian Surveys mg/kg Dry Weight	
	1979	1988	1996	1981	1995
Arsenic	6.7	9.9	11.5		2.3
Cadmium	69	6.9	6.4	35	6.3
Chromium	429	119	103	1,040	319
Copper	602	741	506	870	638
Lead	369	134	111	545	124
Mercury	2.8	5.2	2.1		3.5
Molybdenum	17.7	9.2	15		22
Nickel	135	43	57	160	38
Selenium	7.3	5.2	5.7		3.3
Zinc	1594	1202	830	1,390	823

Sources: Webber and Nichols, 1995; Lue-Hing et al., 1999.

TRACE ELEMENTS IN ANIMALS, HUMANS, SOILS, AND PLANTS

Arsenic (As)

Animals and Humans

Arsenic is toxic to animals and man. The maximum tolerable levels of dietary inorganic As is 50 mg/kg for cattle, sheep, swine, poultry, horse and rabbit. The tolerable level of organic As is 100 mg/kg for the same animals (NRC, 1980). Under natural dietary conditions, As toxicity is uncommon (Gough et al., 1979). There have been reports on cattle and sheep toxicity from grazing on pastures containing high levels of As in soils treated with arsenicals (Selby et al., 1974; Case, 1974). Arsenic bioavailability has been shown to be five times less available than As from the salt Na_2HAsO_4.

Arsenic is believed to be essential to mammals (Chaney, 1983). Several organic arsenic compounds have been fed to pigs and poultry to stimulate growth (Gough et al., 1979). The data cited above indicate that the potential for As toxicity to human and animal food chain from land applied biosolids is very minimal for the following reasons:

- Levels of As in biosolids are very low.
- Arsenic in biosolids is in an organic matrix and is less available than salts; bioavailability of As from an organic matrix is very low.
- The food chain is protected because As phytotoxicity will affect crops consumed by humans and animals
- Arsenic is not readily taken up by plants.

Soils

The two most common inorganic forms of As in soils are arsenate and arsenite. Arsenic under aerobic conditions in the soil reverts to the chemical form of arsenate, which is strongly bound to the clay fraction. This binding reduces the potential of As to migrate through soils and inhibits its uptake by plants. Arsenite is formed under anaerobic conditions and is more phytotoxic. It is not adsorbed on soil particles to as great extent as arsenate. Consequently, more As is in the soil solution and can cause phytotoxicity (Tsutsumi, 1981; Chaney and Ryan, 1994). In flooded soils arsenite will predominate. Phosphate will displace adsorbed As which allows it to leach down and be readsorbed at lower levels (Onken and Hossner, 1995).

Plants

Arsenic is not considered essential to plants and is not readily taken up by plants. It tends to accumulate in the roots, which reduces its concentration in edible above-ground portions of plants (U.S. Department of Agriculture, 1968).

Arsenic can be toxic to plants. The toxicity is a function of the concentration of the soluble, not total, arsenic content of soils (Gough et al., 1979). Toxicity to As has been primarily related to the use of pesticides (Chaney, 1983; Gough et al., 1979). Calcium arsenate, lead arsenate and cupric arsenate (Paris green) were widely used as insecticides (Gough et al., 1979). The use of As insecticides in orchards has resulted in high levels of soluble As, rendering the soils of some orchards unproductive (Gough et al., 1979).

Arsenicals have been used as defoliants in cotton and potatoes (Woolson, 1983). Wells and Gilmore (1977) reported that phytotoxicity to rice occurred when cotton fields were used for rice production. Rice grown in flooded soils is the most sensitive crop to As toxicity from soil As. High concentrations of soil As can be phytotoxic to many crops including peas, potatoes, cotton and soybeans (Stevens et al., 1972; Deuel and Swoboda, 1972). Duel and Swoboda reported that 4.4 µg/g or greater As concentration in cotton and 1 µg/g and greater in soybeans limited yield. Under flooded conditions, the rate of As uptake by rice increased as the rate of plant growth increased.

Jacobs et al. (1970) showed that As residues in soils from potato cultivation, where Na-arsenite was used as a defoliant, decreased yields of vegetables. Stevens et al. (1972) reported that on As contaminated sand, arsenic levels were contained in the potato peel with very low amounts in the tuber.

Most of the data on As toxicity to plants are from the use of salts and not from As in biosolids or other organic matrices. As is phytotoxic before crops can accumulate As to a level which is toxic to humans. Therefore, the food chain is protected (Chaney, 1983).

Cadmium (Cd)

In addition to being a natural element in soils and geological material, Cd enters our environment from fertilizers, phosphatic materials, zinc-associated compounds, plastics, batteries, land application wastes or waste products, coated metals, paints and smeltering and purification of metal ores. Many of the world's agricultural areas are contaminated to some extent with Cd. The use of biosolids could further add to the soil burden.

How significantly could the addition of Cd, through the application of biosolids, impact the food chain and how might this affect the health of animals and humans? The answer to this question depends on numerous factors, including uptake and accumulation by plants and their organs, bioavailability to animals and humans, interrelationship of Cd to other elements related to growth and nutrition, accumulation in organs in relation to age of humans, and diet.

Considerable research has been conducted on Cd in biosolids and potential health impacts. This section highlights some of the key aspects. For greater details, the author encourages readers to explore works by Ryan et al. (1982); Friberg et al. (1974) and Elinder (1985).

Animals and Humans

Cadmium is not considered essential to animals and man. However, limited data suggest the contrary — that this element may be essential (NRC, 1980). Cadmium is toxic to animals and man. It is retained in the kidney and liver and is probably related to the metal binding protein metallothionein (Kagi and Vallee, 1960).

Acute health effects due to high exposure can result in severe damage to several organs. The data are primarily from experiments with animals and occupational exposure. Cd exposure in fumes (e.g., in plating operations) can result in pulmonary edema. Lucas et al. (1980) reported on lethal effects of Cd fumes. Acute symptoms by Cd fumes occur after 4 to 6 hours of exposure and include cough, shortness of breath and tightness of chest. Pulmonary edema may appear within 24 hours, often followed with bronchopneumonia (Ryan et al., 1982).

The accumulation of CD in the body appears to increase up to the age of 50 and then decreases (Elinder et al., 1976). It has been estimated that the half-life of Cd in the kidney ranges from 18 to 33 years (NRC, 1980). Chronic health effects are principally manifested in the kidney. Other chronic health effects believed to be related are; hypertension, respiratory effects, carbohydrate metabolism, carcinogenesis, teratogenesis and damage to liver and testicles.

Scientists disagree about the effects of cadmium on cancer. After reviewing the literature, Fasset (1975) states that the evidence for carcinogenesis appears to be doubtful. Sunderman (1971, 1978) also found the evidence on cancer to be meager. Kolonel (1976) compared 64 cases of renal cancer in white males with controls and indicated significant association of renal cancer to exposure to cadmium. Several authors indicate a relationship between the formation of Leydig-cell tumors in testes of animals (Reddy et al., 1973; Levy et al., 1973; Malcolm, 1972).

Adsorbed Cd is bound to a low-molecular-weight protein to form metallothionein, which accumulates in the kidney cortex (Chaney, 1983). Also, Cd apparently competes with Zn on the same binding sites, presumably thiol groups (Pulido et al.,1966). Renal chronic effects are manifested by proteinuria and tubular dysfunction. Friberg et al. (1974) estimated that the critical level of damage in the renal cortex is 200 µg/g wet weight.

Other than occupational exposure, the intake of Cd is principally from food and water and, in the case of smokers, from smoking. Gastrointestinal adsorption is poor. It is estimated that approximately 5% of the intake of Cd is adsorbed through the gut (WHO, 1982; Shaikh and Smith, 1980).

The tobacco plant accumulates Cd in the leaves as a result of its presence in the soil and concentrations can range from 1 to 6 µg/g. The primary source is from phosphate fertilizers. Furthermore, tobacco is grown on acidic soils, which enhance the availability and plant uptake of Cd. Each cigarette can contain from 1.2 to 2.0 µg/g Cd. Cigarette smoke can be a very significant source of Cd to the body because adsorption through the lung is high. Friberg et al. (1974) estimated that nearly 50% of the Cd in cigarette smoke is absorbed. Higher values have been suggested. Thus for smokers, more than one-third of the body burden could be from smoking. An individual who smokes one pack of cigarettes per day could receive about one-half of the body burden of Cd from this source. Sharma et al. (1983) demonstrated that cigarette smoking had a more pronounced and significant effect on whole blood Cd levels than intake from ingestion of oysters that have high Cd concentrations. Table 4.7 shows the potential intake of Cd from various sources.

Table 4.7   U.S. Daily Intake and Retention of Cadmium from Various Sources

Source	Concentration	Intake µg	Retained [1]µg
Total diet	0.04 µg/g	51	2.30
Drinking water	0.0014 µg/g	2.8	0.13
Air	0.006 µg/m	0.12	0.05
Cigarettes (20)	1.0 µg/g	3.1	1.4

[1] Assuming 4.5% of ingested Cd and 45% of inhaled Cd are retained.
Source: Parr et al., 1977.

The risk reference dose (RfD) for Cd is 70 µg/day. This RfD is designed to protect the highly exposed individuals (Chaney and Ryan, 1993). This level is also the maximum permissible level of dietary Cd established by the World Health Organization.

Daily intake of Cd varies. It has been estimated that the variation ranges from 12 µg/day to 51 µg/day (Braude et al.,1975; Ryan et al., 1982; Chaney and Ryan, 1993).

Although there is no evidence that human exposure to Cd from biosolids applied to land has resulted in health effects, there is strong evidence of adverse health effects from exposure to contaminated foods. A prominent example relating Cd contamination of food crops and water occurred in Japan. In 1955 Drs. Hagino and Kohno (Yamagata and Shigematsu, 1970) reported on a disease they named "Itai-itai" or "ouch-ouch," which was the result of severe bone pains. The disease was manifested by osteomalacia, pathologic features similar to Fanconi's Syndrome and pain in inguinal (groin) and lumbar regions and joints. Other manifestations were proteinuria and glycosuria and an increase of serum alkaline-phosphate and decrease of inorganic phosphorus. Duck gait was evidenced as well as roentgenological appearance of the transformation zone of the bone with proneness to fracture. The affected individuals were primarily childbearing women over 40 years of age.

In 1968, the Japanese Ministry of Health and Welfare reported that the disease was caused by chronic Cd poisoning. Cadmium polluted rice fields were the result of discharges from mine smeltering activities. Inhabitants accumulated Cd from food and water. Yamagata and Shigematsu (1970) found paddy-soil levels of Cd between 2.2 and 7.2 parts per million (ppm) and rice levels between 0.72 and 4.17 ppm as compared with control levels of less than 1 ppm in soil and 0.03 to 0.11 ppm in rice. The latter are relatively low levels in both soils and crops in terms of potential toxic effects on humans. More recent data have shown that the Japanese conditions have been greatly affected by their diet and the bioavailability of Cd. The Japanese consume large quantities of rice. A typical consumption of 300 g per day of rice containing 1 ppm of Cd would result in an addition of 300 µg Cd.

McKenzie and Eyon (1987) and McKenzie et al. (1988) reported that New Zealand adults in a region of that country consumed large quantities of dredge or bluff oyster (*Tiostrea lutaria*) which has a high concentration of Cd. Consumption of Cd from oysters and fecal output of Cd in some New Zealand adults exceeded the fecal output reported by the Japanese. However, the New Zealand residents did not have renal damage or symptoms similar to the Japanese residents. The main differences between the two populations were their diets and the bioavailability of Cd in different foods or diets (Chaney and Ryan, 1993). Other studies in England (Sherlock, 1984) and East Greenland (Hansen et al., 1985), where populations consumed high levels of Cd, did not show adverse health effects.

Fox (1988) indicates that Zn, Fe, Cu, Ca, ascorbic acid and protein may interact with dietary Cd. Iron particularly affects the bioavailability of Cd. Low Fe diets contributed to higher Cd retention (Flanagan et al., 1978). Increased dietary Zn apparently induces biosynthesis of metallothionein, which binds both Cd and Zn (Chaney, 1988). Calcium deficiency also increases Cd adsorption (Chaney, 1988). Thus, the potential effect of Cd from food crops is not only a function of levels in the crops but also Cd bioavailability and human nutrition conditions.

Soil

In soil of nonpolluted areas, Cd is usually less than 1 mg/kg dry weight (Page et al., 1981). However, high levels of Cd have been found in certain areas as a result of geological parent material sources (Lund et al., 1981). As indicated earlier, the extensive evaluation of agricultural soils in the United States by Holmgren et al. (1993) found levels of Cd ranged from <0.0010 to 2.0 mg/kg (Table 4.2). Andersson (1976) reported that, in 361 Swedish soil samples, Cd ranged from <0.063 to 0.249 µg/g. Cd concentrations in soil are dependent on the parent material, secondary material and organic substances (Elinder, 1985). Cd mobility and uptake by plants is affected by soil pH, organic matter, iron and clay.

Plants

Cadmium is not essential to plants and its toxicity is generally moderated (Gough et al., 1979). Depressed growth of plants appears to be when plant tissue Cd exceeds 3 µg/g (Allaway, 1968; Millner et al., 1976). Bingham (1979) indicated that a 25% yield reduction for various crops resulted when Cd concentration ranged from 7 to 160 µg/g dry weight. Uptake, accumulation and translocation of Cd by plants varies considerably (CAST, 1980; Bingham, 1979; Bingham et al., 1975, 1976; Chaney and Hornick, 1978; Chaney, 1983; Dowdy and Larson, 1975). Accumulation varies between plants and within plants. Different organs of the plants studied accumulated Cd to varying degrees. Primarily, Cd accumulates in leaves (Chaney and Hornick, 1978; Bingham et al., 1975, 1976). Detailed information is presented in Chapter 5. Tobacco, as a leafy plant grown on acidic soils, accumulates Cd. Chaney et al. (1978) reported that tobacco grown on biosolid-amended soil accumulated 44 µg/g when the soil contained 1 µg/g of Cd. Smoking is a major source of Cd in the body.

Cadmium uptake by plants is affected by several soil factors including pH, organic matter, soil particle size, chloride concentration, total soil Cd, Zn status, hydrous iron and the presence of manganese and aluminum oxides (Brown et al., 1996; McBride, 1995; Mclaughlin et al., 1994; Corey et al., 1987).

Chromium (Cr)

Animals and Humans

Chromium is essential to animals and man (Underwood, 1977; NRC, 1980). For example, it is necessary for normal glucose metabolism in animals (Van Campen, 1991). Humans are often deficient in this element as a result of low levels in plants. An organic form of the element is a cofactor in insulin response controlling carbohydrate metabolism (Toepfer et al., 1977). Chromium does not appear to concentrate in any specific organ. However, it was found to accumulate in the lung, probably as a result of inhalation of dust containing Cr (Mertz, 1967).

Chromium tends to decline in body tissues with age (Anderson and Koslovsky, 1985). Anderson (1987) indicates that Cr deficiency affects glucose intolerance, elevated serum cholesterol and elevated serum triglycerides. Other manifestations include elevated blood-insulin concentrations, glycosuria, hyperglycemia, neuropathy and encephalopathy. Several foods that are good sources of Cr include brewer's yeast, meat, cheese and whole grains. Chromium as chromium picolinate is sold as a dietary supplement.

Chromium is also toxic. Certain chemical forms have been shown to be mutagenic and carcinogenic. Chromate in dusts and mist has been shown to result in nasal cancer (NRC, 1980). Occupational manifestations of cutaneous and nasal mucous-membrane ulcers and contact dermatitis have been reported following exposure to hexavalent chromium (Gough et al., 1979). Van Campen (1991) indicates that trivalent Cr at the dietary levels normally encountered is not likely to be toxic.

Soils

Chromium exists in soils in low redox forms: chromic (Cr^{3+}) and chromate (Cr^{6+}). Chromate is rapidly reduced to chromic in soils. This reduction occurs more rapidly in acidic soils (Bartlett and Kimble, 1976; Cary et al., 1977). Chromium is strongly adsorbed and chelated by soils. Chromic is insoluble and also strongly sorbed. These reactions reduces its presence in the soil solution and reduces plant uptake. Chromate is rapidly reduced to Cr^{3+} by reaction with organic matter or other reducing agents in soils. However, Bartlett and James (1979) showed that Cr^{3+} can be oxidized to Cr^{6+} by Mn-oxides.

The inert nature of Cr compounds and chelates (slow kinetics of reactions in soils) can be important in limiting the potential for oxidation of applied Cr^{3+} and leaching of Cr^{6+}. Equilibrated Cr^{3+} in soils is essentially inert under the conditions of pH, chelation and redox found in nearly all soil materials. If Cr^{3+} is only sparingly soluble in the soil solution, the oxidation reaction does not proceed. This inert nature is an important source of environmental protection against adverse effects of Cr^{3+} applied to soil by biosolids or other organic amendments (Chaney et al., 1996).

Plants

Chromium is nonessential to plants. It is phytotoxic as chromate (Cr^{6+}). Chromium toxicity varies greatly with species. On some soils high in Cr (e.g., serpentine soils), several species tolerate the high levels of Cr. Chromate is more soluble and more available for plant uptake than Cr^{3+} usually found in biosolids. Chromium has produced toxicity symptoms to tobacco, corn and oat at soil chromate levels of 5 to 16 ppm (NRC, 1974). In tobacco, symptoms occurred when concentrations in leaves ranged from 18 to 24 mg/kg and 375 to 410 mg/kg in roots. Corn leaves exhibited symptoms at 4 to 8 mg/kg; and oats at 252 mg/kg (NRC, 1974).

Cr^{6+} phytotoxicity is manifested by reduced root development. Cumulative application of Cr in biosolids had occurred in the field in the United States at least as high as 300 kg/ha without adverse effects (Chaney et al., 1996). Long-term studies in Minnesota indicated that there was no reduction in corn yield nor in Cr accumulation when 1045 tonnes/ha of Cr were applied through biosolids (Dowdy et al., 1994).

Plant uptake of Cr is very limited because it is reduced in the roots to Cr^{3+} and is not translocated to the above portions of the plant. Even under Cr^{6+} phytotoxicity, the level of Cr is less than 10 mg/kg. Because crops have low Cr levels even if grown on soils very high in Cr, the food chain is protected against excess Cr in plant tissues. Plants grown on serpentine soils containing as much as 1% (10,000 mg Cr/kg) do not exhibit Cr phytotoxicity (Chaney et al., 1996).

Copper (Cu)

Copper has been used for decades as an algecide and fungicide. Bordeaux (a mixture of copper sulfate and lime) has been used as a spray in vineyards and vegetable crops. Copper has also been added as diet additive to swine and poultry and thus excreted in the manure. Industrial pollution also added Cu to soils. Cu deficiencies in agriculture are more common than toxicities. Cu is often added to agricultural crops grown on sandy soils.

Animals and Humans

Copper is essential to animals and man. Cu is associated with Cu proteins and enzymes. Cu appears to be essential for normal reproduction (Underwood, 1981). Copper is also toxic to animals. Toxicity to sheep and cattle has been reported to occur at levels of 25 to 100 mg/kg dry diet (NRC, 1980). Sheep appear to be particularly sensitive to copper. Relatively high concentrations are found in the liver, brain, heart and hair (Miller et al., 1991).

Cu is toxic to man but poisoning is rare. It is concentrated in the liver and depends on age and diet (Van Campen, 1991). Wilson's disease is caused by the buildup of Cu in the liver and central nervous system as a result of the body's inability to excrete it (Scheinberg, 1969). Acute poisoning causes gastrointestinal ulcerations, hepatic necrosis, hemolysis and renal damage (Van Campen, 1991).

Cu deficiencies include anemia associated with Fe adsorption and utilization; bone and cardiovascular disorders, mental and/or nervous system deterioration and defective keratinization of hair (Van Campen, 1991). Oysters, organ meats, mushrooms, nuts and dried legumes are considered a good source of dietary Cu (Van Campen, 1991).

Marston (1950) noted that Cu deficiency in animals inhibited hemoglobin formation. It was found that Cu was not actually part of the hemoglobin molecule, but it performs an important function in the formation of hemoglobin.

Soils

Cu is a very immobile micronutrient in the soil (Moraghan and Mascagni, 1991). Its reaction in soil is similar to several other metals. A low soil pH increases Cu solubility and availability to plants. Soil organic matter binds Cu to form complex chelates (Logan and Chaney, 1983). Optimum acidity for the complex formation of Cu ranges from pH 2.5 to 3.5 for humic acid and pH 6 for fulvic acid. The reaction of Cu (II) with carboxyl groups in humic acid has been suggested by Schnitzer (1978) and Boyd et al. (1981). Chaney and Giordano (1977) cited several cases of reversion of Cu to unavailable forms. Organic soils such as peats and mucks generally have low available Cu or the Cu is complexed resulting in crop deficiencies.

Plants

Copper in the water-soluble and exchangeable forms is considered available to plants (Shuman, 1991). The normal range of Cu in plants is from 5 to 20 mg/kg in tissues. Phytotoxicity occurs in most plants at about 25 to 40 mg/kg dry foliage (Chaney and Giordano, 1977; Page, 1974).

Cu toxicity is manifested by dark green leaves followed by induced Fe chlorosis, thick, short, or barbed-wire looking roots and depressed tillering (Jones, 1991). Toxic levels of Cu in plants are dependent on the concentration of clay and organic matter.

Cu is contained in enzymes and plant proteins. It plays an important part in photosynthesis and respiration. In many species Cu concentrations of less than 5 mg/kg are indicative of deficiency. Cu deficiency results in depressed growth and reproduction (i.e., formation of seeds and fruits). Deficiency symptoms are chlorosis (white tip, reclamation disease), necrosis, leaf distortion and dieback. The most noticed symptoms of Cu deficiency are reduced seed and fruit production as a result of male sterility (Romheld and Marschner, 1991). Phosphate, manganese, or zinc may directly compete with available Cu, potentially resulting in Cu deficiency. Other factors that could affect Cu concentration in plants are microbial activity, moisture, pH, redox potential and plant species.

Lead (Pb)

Animals and Humans

Lead is a nonessential element to humans, animals and plants. It is toxic to humans and in the past several decades there has been an increased concern for its toxicity to children. There are many sources of Pb exposure to humans. Air, water, dust, soil and diet are the primary sources.

The acceptable blood levels have become more restrictive over the years. In 1985 the acceptable blood level was 25 µg/dl, whereas in 1991, it was reduced to 10 µg/dl (CDC, 1991). Its toxicity is based on body weight. Generally, toxic intake is 1 mg, lethal intake is 10 g (Pais and Jones, 1997). Lead toxicity to many animals occurs at about 30 mg Pb/kg diet (NRC,1980). Dietary Fe, Ca, Zn, P and fiber interact with Pb (Mahaffey and Vanderveen, 1979; Levender, 1979).

The primary risk to animals and man is from ingestion of particles with high concentration of Pb. U.S Environmental Protection Agency, in conducting risk assessment on land application of biosolids, determined that soil ingested by children represents the highest level of risk. Soil Pb can result in excessive blood Pb if it is somewhat bioavailable to humans (Chaney, 1983). Adsorption of dietary Pb, especially the inorganic forms, is decreased by increasing dietary levels of Ca, P, Fe and Zn. In ruminants, tolerance of dietary Pb is increased by added levels of dietary sulfur or sulfate. Se sometimes also provides protection.

Soils

Soils throughout the world contain Pb. Some of the Pb in soils is from natural geological sources whereas other soils have become polluted or contaminated by man. Soils throughout the world have been polluted and contaminated with Pb from leaded gasoline, paints and emission sources. The past use of organic Pb additives to gasoline is the major source of Pb contamination of surface soils.

Geochemical concentrations of lead (Pb) in the earth's crust have been estimated at 12.5 mg/kg (*CRC Handbook of Chemistry and Physics,* 1990). Pb concentration in natural soils was reported by Conner and Shacklett (1975) to range from 2.6 to 25 mg/kg and Holmgren et al. (1993) reported that Pb in agricultural soils range from <1 to 135 mg/kg with a mean of 10.5 mg/kg. Pais and Jones (1997) indicated that the total content of Pb in soils ranged from 3 to 189 mg/kg, and the natural background level ranged from 10 to 67 mg/kg.

Pb does not readily move through the soil profile. Numerous studies have shown that it remains on or near the soil surface (Chaney et al., 1988). This feature has both positive and negative attributes. By remaining at the surface and not moving through the soil profile the potential for contamination of ground water resources and drinking water is very low. However by remaining at the surface, the potential exposure to children by ingestion of soils or by dust is much greater. Lead in surface soils stems primarily from the use of Pb additives to gasoline and atmospheric deposition from industrial sources. Surfaces painted with lead-based paint also can be an important source of soil Pb.

The reaction of Pb in soil is affected by adsorption by the soil cation exchange complex; precipitation by sparingly soluble compounds; and formation of relatively stable complex ions or chelates as a result of interaction with the soil organic matter. The degree of sorption will depend on the soil's electronegativity and the ionization potential of the adsorbed ions as well as the exchange complex. Lead solubility in soil decreases as the soil pH increases. In noncalcareous soils, the solubility of Pb appears to be related to the formation of hydroxides and phosphates [e.g., $Pb(OH)_2$, $Pb_3(PO_4)_2$]. In calcareous soils, lead carbonate ($PbCO_3$) appears to be dominant.

Lead combines with ligands to form stable metal complexes and chelates and in this matter is complexed with soil organic matter. In a letter to the USDA, Chaney suggested that the use of biosolids products in urban areas where Pb concentration from automobile emissions is high could reduce its availability to children ingesting soil.

Plants

Lead is not essential to plants. Phytotoxicity to plants is rare although elevated levels in plants have been reported. Plants growing in soils in lead-rich sites absorbed large amounts of Pb without exhibiting phytotoxicity (Shackelette, 1960). On lead-mined soils, high levels have been found in grasses (Johnson and Proctor, 1977). Baumhardt and Welch (1972) applied up to 3200 kg/ha of Pb to a soil with a pH of 5.9 and did not observe a decrease in corn yield.

Plants can take up Pb into the roots; however little is generally translocated to the upper portion of the plants because insoluble Pb-phosphate is formed in the roots. As a result, the food chain is protected from excess lead.

Mercury (Hg)

Mercury is nonessential to plants, animals and humans. It has been widely used as a fungicide in agriculture and horticulture. Hg compounds inhibit bacterial growth and have been used as antiseptics and disinfectants (Gough et al., 1979).

Animals and Humans

Mercury is toxic and is not essential to animals and man. Methylated mercury is the most toxic form. Methyl mercury (MeHgOH) is a neurotoxin (WHO, 1990). Mercury poisoning of humans from fish and contaminated seed have been reported (Friberg and Vostal, 1972; Bakir et al., 1973)

Soils

Mercury reaches the soil from atmospheric deposition, use of fungicides or insecticides and wastes. Once in the soil, Hg reacts with the exchange complex of the clay and organic fractions, forming both ionic and covalent bonds. It can be precipitated as insoluble phosphate, carbonates and sulfides. Mercury can be chelated to organic matter. The binding of Hg to organic matter and clay reduces its potential for groundwater contamination. Many mercurial compounds, both organic and inorganic, can be volatilized, converted to sulfur and chloride compounds, or adsorbed by sesquioxide surfaces. Oxidized Hg can be reduced to volatile Hg^0 in the upper surface layer of soil (Carpi and Lindberg, 1997). Sunlight, surface soil temperature and soil moisture are factors that affect Hg emissions (Carpi et al., 1997). Liming a soil to a pH exceeding 6.5 reduces its availability to plants.

Plants

Mercury can enter plants through the roots and leaves from foliar sprays, dusts, or vapors. Mercury in plant tissues is principally organic and primarily as methyl mercury. It is soluble in water and is available for incorporation into tissues of aquatic organisms (Gough et al., 1979). Toxicity to greenhouse plants from volatilized elemental Hg has been observed (Shacklett, 1970).

Molybdenum (Mo)

Animals and Humans

Molybdenum is essential to animals and man. It functions as molybdoflovoprotein in maintenance of levels of xanthine oxides. Deficiencies have been reported in sheep (Underwood, 1977; NRC, 1980). Molybdenum is toxic to animals — especially ruminants. Monogastric animals are more tolerant of Mo than ruminants. In sheep, the highest concentration of Mo accumulates in the liver, followed by the kidney and then the lung (Grace and Martinson, 1985).

Molybdenosis occurs as a result of induced Cu deficiency as well as excess Mo and low sulfate-Se concentrations in forage (Mills and Davis, 1987; Underwood, 1977; Gough et al., 1979). This disease is also referred to as teart disease or peat scours. Mo supplements can alleviate chronic Cu toxicity in sheep (NRC, 1980). Mo in animals is also important in enzyme metabolism (Miller et al., 1991). Both increased growth (Payne, 1977) and decreased growth (Anke et al., 1985) have been reported. Yang et al.,(1985) indicated that Mo in drinking water inhibited mammary carcinogenesis.

Molybdenum toxicity in humans is rare. Gough et al. (1979) cite a case in India where individuals consumed sorghum that was grown on alkali soils high in Mo. The peasants developed a crippling syndrome of knock-knees (*genu valgum*).

Soils

Molybdenum is associated with both the organic and inorganic soil fractions and is often found with iron oxides. It is sorbed by sesquioxides (especially iron) and clay minerals. The amount of Mo absorbed by clays and hydrous oxides decreases with increasing soil pH to about 7.5, after which limited absorption takes place. Increasing the soil pH by liming from pH 5 to pH 7 increased Mo uptake by plants.

Molybdenum is present in soils as anionic molybdate. Soils adsorb Mo strongly under acidic conditions (Chaney, 1983). Therefore under acidic conditions Mo does not readily leach through the soil. In acid soils Fe-molybdates are the predominant form of inorganically combined Mo. Iron oxides and P in soils greatly affect Mo sorption. Large quantities of P in the soil will replace Mo, which is bound to iron oxides. Maximum sorption of Mo occurs in soils at pH 4.2 (Cast, 1976).

In well-aerated soils, the predominant chemical forms are MoO_4^{-2} and HMO_4^{-1}. Under alkaline conditions, Mo is more readily taken up by plants and also can leach through the soil. Several soils are high in Mo resulting in molybdenosis of sheep and cattle (Kabota, 1977; Gupta et al., 1978). Molybdenosis has been found in Kern County, California, the San Joaquin Valley of California, Nevada and Oregon (Barshad, 1948; Kabota et al., 1961, 1967).

Plants

Molybdenum is an essential trace element for plants. It is contained in several enzymes, including sulfite oxidase, aldehyde oxidase, and xanthine dehydrogenase. It is also present in the enzyme nitrogenase, which is responsible for molecular nitrogen formation and in the nitrate reductase enzyme, which is responsible for nitrification. Plants suffering from Mo deficiency often exhibit signs of N deficiency.

Plants can tolerate high levels of Mo and translocate it to the edible portions. Molybdenum content of plants varies. Legumes accumulate more Mo than grasses (Allaway, 1977; Kabota, 1977). Forage plants have been shown to contain 200 mg/kg Mo, which is more than 600 times the amount necessary for growth (Evans et al., 1950). Vegetables can also accumulate Mo (Gupta ct al., 1978). The Mo availability to plants is closely related to the kinds and amounts of Fe compounds in the soil.

Gerloff et al. (1959) and Olsen and Watanbe (1979) indicated that in alkaline and calcareous soils of semiarid regions, high Mo concentrations may result in plant induced Fe deficiency indicated by chlorosis. Molybdenum accentuated Fe deficiency at low levels of available Fe. Hence this interaction of Fe and Mo may be significant in alkaline soils where pH impacts Fe availability and native soils are high in Mo.

Molybdenum deficiency symptoms manifest in malformation of leaves (whiptail), interveinal mottling and marginal chlorosis of older leaves, followed by necrotic spots at leaf tips and margins (Romheld and Marschner, 1991). Mo deficiency results in destruction of embryonic tissues. Deficiency generally occurs in various plant tissues when concentrations are in the range of 0.03 to 0.15 mg/kg.

Molybdenum toxicity manifests itself through yellowing or browning of leaves and depressed tillering. Toxicity generally occurs when plant tissue of mature leaves is >100 mg/kg (Jones, 1991). Molybdenum availability to plants is primarily influenced by the level of Mo, sulfate, phosphate, pH, and the nature and content of free Fe-oxides.

Nickel (Ni)

Nickel is essential to plants, animals and man. The greatest concern in biosolids relates to potential phytotoxicity. Ni is ubiquitous in the environment. Application of phosphorus fertilizers, coal, fly ash and biosolids can increase soil Ni and result in an increase in plant uptake.

Animals and Humans

Ni is an essential micronutrient for animals (Nielsen, 1984). However, deficiency has seldom been observed. Ni is toxic to animals and occurs at 50 to 100 mg Ni/kg feed where Ni is added as soluble salts (NRC, 1980). Its toxicity to mammals is low, possibly as a result of limited absorption (Gough et al., 1979). Animals appear to have a higher tolerance to Ni than do plants. Therefore there appears to be little possibility of Ni toxicity to animals consuming Ni at levels that are phytotoxic to plants.

The levels in biosolids do not appear to endanger animals consuming crops grown on biosolids-applied land. Phytotoxicity would reduce plant growth prior to reaching Ni levels that could be harmful to animals and man. Toxicity to humans is low. Occupational pathogenic evidence has been observed.

Soil

Nickel is found in all soils. There is a great variation in Ni content of soils, ranging from 5 to 500 mg/kg (Swaine, 1955). Soils that contain the mineral serpentine may have levels up to 500 mg/kg (Asher, 1991). Soil can be contaminated with nickel from superphosphate, automobile exhausts, industrial smokestacks, smelting and biosolids. The soil chemistry of nickel is similar to that of many metals. Factors such as cation exchange capacity (CEC), organic matter, chelation, pH, solubility and precipitation are important in determining the availability of Ni to plants. The concentration of extractable Ni in soil appears to be governed by the Fe and Mn hydrous oxide surfaces that act as a sink for Ni and by organic chelates that complex Ni (Cast, 1976).

Plants

Ni was shown to be a trace constituent of plants and beneficial since the mid 1920s (Asher, 1991). Its essentiality was demonstrated in 1987 when (Brown et al.,1987a; 1987b) described deficiency symptoms in wheat, oats and barley. These included interveinal chlorosis, premature senescence and inability of the leaves to completely unfold. Ni appears to be important in urea metabolism and involved in enzymatic reactions.

Ni toxicity manifests itself in chlorosis and yield reduction. Cunningham et al. (1975) found that increasing Ni content in biosolids significantly increased tissue Ni content. However no toxicity was observed when 81 mg/kg in biosolids and 100 mg/kg Ni as $NiSO_4$ were applied to soil. Phytotoxicity manifests in interveinal chlorosis of young leaves at about 50 to 100 mg/kg dry weight (Chaney, 1983). Some plants are tolerant of Ni and are Ni accumulators, where concentration can exceed 1000 mg/kg (Asher, 1991).

Selenium (Se)

Se is both beneficial and toxic to plants, animals and humans (McNeal and Balistrieri, 1989). Se is found in virtually all geological materials on earth and is a component of soils due to weathering of rocks. It enters the atmosphere and the environment from volcanic activity and burning of fossil fuel, especially coal (McNeal and Balistrieri, 1989). Several areas in the United States contain high Se concentrations in soils, resulting in Se toxicity to cattle. These areas are mostly in western states and are associated with known seleniferous geological formations (Boon, 1989).

The application of biosolids to agricultural soils did not increase Se uptake by plants (Dowdy et al., 1994; Logan et al., 1987).

Soil

The concentration of Se in most soils ranges from 0.01 to 2 mg/kg. There are, however, areas in the world where reported values exceed 1200 mg/kg (Swaine, 1955). Soil pH and redox potential affect the forms of Se (Chaney, 1983; Elrashidi, et al., 1989). Selenate is the major species in the soil solution and therefore is readily absorbed by plants. Selinite is adsorbed on soil and is only marginally available to plants.

Plants

Se is not considered essential for all plants (Gough et al., 1979). Plants grown on seleniferous soils can absorb large amounts of Se. Mayland et al. (1989) reported that certain species can have concentrations of Se in the hundreds or thousands of mg/kg. Resistance to Se toxicity varies considerably so that general toxicity levels cannot be reliably estimated (Gough et al., 1970).

Zinc (Zn)

Animals and Humans

Zn is essential to animals and humans and is an indispensable component of more than 200 enzymes and proteins (Hambridge et al., 1987). There have been several examples of Zn deficiency reported (Prasad et al., 1963; Hambridge et al., 1972).

Zinc is a requirement for all species of animals that have been examined. Requirements for young domestic animals and poultry range from 40 to 100 mg/kg in natural diets (NRC, 1980). Higher levels are required when diets contain excessive calcium and especially when also combined with vegetable proteins and excessive phytate. Zinc requirements and tolerance to Zn are affected by several nutrients including vitamin A and D, Cu, Mn, Fe, Pb and Cd.

Soil

The primary natural source of Zn in soils is from weathering of the ferromagnesian minerals and sphalerite (Chesworth, 1991). Zn is also introduced to agricultural soils from phosphate fertilizers as well as atmospheric deposition. Soil clays, organic matter and hydrous oxides play a major role in the retention and adsorption of Zn (Harter, 1991).

Plants

Zinc is essential and can also be toxic to plants. Staker and Cummings (1941) reported toxicity on spinach, lettuce and carrots grown on peat soils in New York. The concentration of Zn in plant species varies widely. Zn deficiency in plants as a result of low Zn in soil occurs in many of the cropping areas of the United States and throughout the world (Bould et al., 1984; Welch et al., 1991). Chapman (1966) indicated that Zn deficiencies in plants occurred when Zn levels were less than 20 to 25 mg/kg dry matter. Usually Zn deficiency in plants is associated with high-pH soils or with coarse-textured, highly leached acid soils (Welch et al., 1991).

CONCLUSION

Many of the regulated heavy metals are essential to animal and human health and plant growth. An element such as zinc is often deficient in soils and, as a result, affects crop production. This same element is vital to human development. The use of biosolids can often alleviate plant micronutrient deficiencies.

Other heavy metals or trace elements are toxic to animals, humans and plants. Their addition to soils through land application should be minimized. As indicated in Chapter 1, due to industrial pretreatment during the past decade, heavy metals have been significantly reduced in biosolids.

REFERENCES

Allaway, W.H., 1968, Agronomic controls over the environmental cycling of trace elements, pp. 235–274. A.G. Norman (Ed.), *Advances in Agronomy*, vol. 20, Academic Press, New York.

Allaway, W.H., 1977, Perspectives on molybdenum in soils and plants, pp. 317–339, W.R. Chappell and K.K. Petersen (Ed.), *Molybdenum in the Environment*, Marcel Dekker, New York.

Anderson, R.A., 1987, Chromium, pp. 225–224. W. Mertz (Ed.), *Trace Elements in Human and Animal Nutrition*, 5th ed., vol. 1, Academic Press, San Diego, CA.

Anderson, R.A. and A. Koslovsky, 1985, Chromium intake, absorption and excretion of subjects consuming self-selected diets, *Am. J. Clin. Nutr.* 41: 1177–1183.

Andersson, A. and K.O. Nilsson, 1976, Influence on the levels of heavy metals in soil and plant from sewage sludge used as a fertilizer, *Swedish J. Res.* 6: 151–159.

Anke, M., B. Groppel and M. Grun, 1985, Essentiality, toxicity, requirements and supply of molybdenum in humans and animals, pp. 154–157. C.F. Mills et al., (Ed.), *Trace Elements in Man and Animals* – TEMA5, Commonwealth Agric. Bureaux, Farnham, England.

Asher, C.J., 1991, Beneficial elements and functional nutrients, pp. 703–723, J.J. Mortvedt, F.R. Cox, L.M. Shuman and R.M. Welch (Eds.), *Micronutrients in Agriculture*, Soil Science Society of America, Madison, WI.

Ashworth, W., 1991, *The Encyclopedia of Environmental Studies*, Facts on File, New York.

Bakir, K., S.F. Damluji, L. Amin–Zaki, M. Murtadha, A. Khalidi, N.Y. Al–Rawi, S. Tikriti, H.I. Dhahir, T.W. Clarkson, J.C. Smith and R.A. Doherty, 1973, Methylmercury poisoning in Iraq, *Science* 181: 230–241.

Barshad, I., 1948, Molybdenum content of pasture plants in relation to toxicity to cattle, *Soil Sci.* 6: 187–195.

Bartlett, R.J. and J.M. Kimble, 1976, Behavior of chromium in soils: II, Hexavalent forms, *J. Environ. Qual.* 5: 383–386.

Bartlett, R.L. and B. James, 1979, Behavior of chromium in soils: III, Oxidation, *J. Environ. Qual*, 8: 31–35.

Baumhardt, G.R. and Welch, L.F., 1972, Lead uptake and corn growth with soil applied lead, *J. Environ. Qual.* 1: 92–94.

Berthelsen, B.O., E. Steinnes, W. Solberg and L. Jingsen, 1995, Heavy metal concentrations in plants in relation to atmospheric heavy metal deposition, *J. Environ. Qual.* 24: 1018–1026.

Bingham, F.T., 1979, Bioavailability of Cd to food crops in relation to heavy metal content of sludge–amended soil, *Environ. Health Perspect.* 28: 39–43.

Bingham, F.T., A.L. Page, R.J. Mahler and T.J. Ganje, 1975, Growth and cadmium accumulation of plants grown on a soil treated with cadmium–enriched sewage sludge, *J. Environ. Qual.* 4: 207–211.

Bingham, F.T., A.L. Page, R.J. Mahler, and T.J. Ganje, 1976, Yield and cadmium accumulation of forage species in relation to cadmium content of sludge–amended species, *J. Environ. Qual.* 5: 57–60.

Bingham, F.T., A.L. Page, R.J. Mahler, and T.J. Ganje, 1976, Cadmium availability to rice in sludge–amended soil under "flood" and "nonflood" culture, *Soil Sci. Soc. Am. J.* 40: 715–719.

Boon, D.Y., 1989, Potential selenium problems in Great Plains soils, pp. 107–121, L.W. Jacobs (Ed.), *Selenium in Agriculture and the Environment*, American Society of Agronomy, Madison, WI.

Bould, C., E.J. Hewitt and P. Needham, 1984, *Diagnosis of Mineral Disorders in Plants*, Chemical Publ., New York.

Bowhay, D. 1997, Screening survey for metals in fertilizers and industrial by–product fertilizers in Washington State, Dept. of Ecology, Ecology Publ. 97–341, Olympia, Washington.

Boyd, S.A., L.E. Sommers and D.W. Nelson, 1981, Copper (II) and iron (III) complexion by the carboxylate group of humic acid, *Soil Sci. Soc. Am. J.* 45: 1241–1242.

Braude, G.L., C. Jelinek, and P. Corneliussen, 1975, FDA's overview of the potential health hazards associated with land application of municipal wastewater sludge, *Proc. Second National Conf. on Municipal Sludge Management and Disposal,* Information Transfer, Inc., Rockville, MD, 214–217.

Brown, P.H., R.M. Welch and E.E. Cary, 1987a, Nickel: a micronutrient essential for higher plants, *Plant Physiol.* 85: 801–803.

Brown, P.H., Welch, R.M., Cary, E.E. and Checkai, R.T., 1987b, Beneficial effects of nickel on plant growth, *J. Plant Nutr.* 10: 2125–2135.

Brown, S.L., R.L. Chaney, C.A. Lloyd, J.S. Angle, and J.A. Ryan, 1996, Relative uptake of cadmium by garden vegetables and fruits grown on long–term biosolid–amended soils, *Environ. Sci. Technol.* 30: 3508–3511.

Carpi, A. and S.E. Lindberg, 1997, The sunlight mediated emission of elemental mercury from soil amended with municipal sewage sludge, *Environ. Sci. Technol.* 31: 2085–2091.

Carpi, A., S.E. Lindberg, E.M. Prestbo and N.S. Bloom, 1997, Methyl mercury contamination and emission to the atmosphere from soil amended with municipal sewage sludge, *J. Environ. Qual.* 26: 1650–1655.

Cary, E.E., W.H. Allaway and O.E. Olson, 1977, Control of chromium concentration in food plants, II: Chemistry of chromium in soils and its availability to plants, *J. Agr. Food Chem.* 25: 305–309.

Case, A.A., 1974, Toxicity of various chemical agents to sheep, *J. Am. Vet. Med. Assoc.* 164: 277–283.

CAST, 1976, Application of sewage sludge to cropland: Appraisal of potential hazards of the heavy metals to plants and animals, Council for Agricultural Science and Technology, Report 64, Ames, IA.

CAST, 1980, Effects of sewage sludge on the cadmium and zinc content of crops, Council for Agricultural Science and Technology, Report No. 83, Ames, IA.

CDC, 1991, Preventing lead poisoning in young children: A statement by the Centers for Disease Control, U.S. Department of Health and Human Services, Atlanta, GA.

Chaney, R.L., 1983, Potential effects of waste constituents on the food chain, pp. 152–240, in J.F. Parr, P.B. Marsh and J.M. Kla (Eds.), *Land Treatment of Hazardous Wastes*, Noyes Data Corp., Park Ridge, NJ.

Chaney, R.L., 1988a, Metal speciation and interactions among elements affect trace element transfer in agriculture and environmental food–chains, pp. 219–260, J.R. Kramer and H.E. Allen (Eds.), *Metal Speciation: Theory, Analysis and Application*, Lewis Publishers, Chelsea, MI.

Chaney, R.L., 1988b, Effective utilization of sewage sludge on cropland in the United States and toxicological considerations for land application. *Proc. 2nd. International Symposium on Land Application of Sewage Sludge,* Tokyo, Japan, Association for the Utilization of Sewage Sludge.

Chaney, R.L. and P.M. Giordano, 1977, Microelements as related to plant deficiencies and toxicities, pp. 234–279, L.F. Elliot and F.J. Stevenson (Eds.), *Soils for Management of Organic Wastes and Waste Waters*, American Society of Agronomy, Madison, WI.

Chaney, R.L. and S.B. Hornick, 1978, Accumulation and effects of cadmium on crops, *Proc. First International Cadmium Conference,* Metals Bulletin Ltd., London, England, pp. 125–140.

Chaney, R.L. and J.M. Ryan, 1993, Heavy metals and toxic organic pollutants in MSW composts: Research results on phytoavailability, bioavailability, fate, etc., H.A.J. Hoitink and H.M. Keener (Eds.), *Science and Engineering of Composting: Design, Environmental, Microbiological and Utilization Aspects*, Renaissance, Worthington, OH.

Chaney, R.L. and J.A. Ryan, 1994, Risk based standards for arsenic, lead and cadmium in urban soils, pp. 1–130, G. Kreysa and J. Wiesner (Eds.), *Die Deutsche Bibliothek – CIP–Einheitsaufnahme*, DECHEMA, Frankfurt.

Chaney, R.L., P.T. Hundemann, W.T. Palmer, R.J. Small, M.C. White and A.M. Decker, 1978, Plant accumulation of heavy metals and phytotoxicity resulting from utilization of sewage sludge and sludge compost on cropland, *Proc. National Conference on Composting Municipal Residues and Sludges,* Information Transfer, Inc., Rockville, MD, pp. 86–96.

Chaney, R.L., J.A. Ryan and S.L. Brown, 1996, Development of the US–EPA limits for chromium in land–applied biosolids and applicability of these limits to tannery by–product derived fertilizers and other Cr–rich soil amendments, *Proc. Chromium Environmental Issues Workshop,* San Miniato, Italy.

Chapman, H.D.,1966, Diagnostic criteria for plants and soils, Div. Agric. Sci., University of California, Riverside.

Chesworth, W., 1991, Geochemistry of micronutrients, pp. 1–30, in J.J. Mortvedt, F.R. Cox, L.M. Shuman and R.M. Welch (Eds.), *Micronutrients in Agriculture,* 2nd ed., Soil Science Society of America, Madison, WI.

Connor, J.J. and H.T. Shacklette, 1975, Background geochemistry of some rocks, soils, plants and vegetables in the conterminous United States, U.S. Geological Survey, Washington, D.C.

Corey, R.B., L.D. King, C. Lue–Hing, D.S. Fanning, J.J. Street, and J.M. Walker, 1987, Effects of sludge properties on accumulation of trace elements by crops, pp. 25–51, A.L. Page, T.J. Logan and J.A. Ryan (Eds.), *Land Application of Sludge — Food Chain Implications,* Lewis Publishers, Chelsea, MI.

Cunningham, J.D., J.A. Ryan and D.R. Keeney, 1975, Phytotoxicity and metal uptake of metal added to soils as inorganic salts or in sewage sludge, *J. Environ. Qual.* 4: 460–462.

Deuel, L.E. and A.P. Swoboda, 1972, Arsenic toxicity to cotton and soybeans, *J. Environ. Qual.* 1: 317–320.

Dowdy, R.H. and W.E. Larson, 1975, The availability of sludge–borne metals to various vegetable crops, *J. Environ. Qual.* 4: 278–282.

Dowdy, R.H., S.E. Clapp, D.R. Linden, W.E. Larson, T.R. Halbach and R.C. Polta, 1994, Twenty years of trace metal partitioning on the Rosemont sewage sludge watershed, C.E. Clapp, W.E. Larson and R.H. Dowdy (Eds.), *Sewage Sludge: Land Utilization and the Environment,* American Society of Agronomy, Madison, WI.

Dudas, M.J. and S. Pawluk, 1980, Natural abundances and mineralogical partitioning of trace elements in selected Alberta soils, *Can. J. Soil Sci.* 60: 763–771.

Elinder, C.–G., T. Kjellstrom, L. Friberg, B. Lind and L. Linnman,1976, Cadmium in kidney cortex, liver and pancreas from Swedish autopsies, *Arch. Environ. Health* 31: 293–302.

Elinder, C.–G., 1985, Cadmium: Uses, occurrence and intake, pp. 23–63, L. Friberg, C.–G. Elinder, T. Kjellstrom and G.F. Nordberg (Eds.), *Cadmium and Health: A Toxicological and Epidemiological Appraisal,* Vol. I, CRC, Boca Raton, FL.

Elrashidi, M.A., D.C. Adriano and W.L Lindsay, 1989, Solubility, speciation and transformations of selenium in soils, pp. 51–63, in L.W. Jacobs (Ed.), *Selenium in Agriculture and the Environment,* American Society of Agronomy, Madison, WI.

Epstein, E., 1997, *The Science of Composting,* Technomic, Lancaster, PA.

Evans, H.J., E.R. Puris and F.E. Bear, 1950, Molybdenum nutrition of alfalfa, *Plant Physiol.* 25: 555–566.

Fasset, D.W., 1975, Cadmium: Biological effects and occurrence in the environment. *Am. Rev. Pharmacol.* 15: 425–535.

Flanagan, P.R., J.S. McLellan, J. Haist, M.G. Cherian, M.J. Chamberlain, and L.S. Valberg, 1978, Increased dietary cadmium absorption in mice and human subjects with iron deficiency, *Gastroenterology* 74: 841–846.

Fox, M.R.S., 1988, Nutritional factors that may influence bioavailability of cadmium, *J. Environ. Qual.* 17: 175–180.

Friberg, L. and J. Vostal. 1972, *Mercury in the Environment*, CRC, Cleveland, OH.

Friberg, L., C.–G. Elinder, T. Kjellstrom and G.F. Nordberg, 1974, *Cadmium and Health: A Toxicological and Epidemiological Appraisal*, Vol. I, *Exposure Dose and Metabolism*, CRC, Boca Raton, FL.

Friberg, L., C.–G. Elinder, T. Kjellstrom and G.F. Nordberg, 1974, *Cadmium and Health: A Toxicological and Epidemiological Appraisal*, Vol. II, *Effects and Response*, CRC, Boca Raton, FL.

Friberg, L., M. Piscator, G. Nordberg, and T. Kjellstrom, 1974, 248. *Cadmium in the Environment*, 2nd ed., CRC, Cleveland, OH.

Gerloff, G. C., P.R. Stout and L.P. Jones, 1959, Molybdenum–manganese–iron antagonisms in the nutrition of tomato plants, *Plant Physiol.* 34: 608–613.

Gough, L.P., H.T. Shacklette, and A.A. Case, 1979, Element concentrations toxic to plants, animals and man, U.S. Department of Interior, Geological Survey, *Geological Survey Bulletin*, 1466, Washington, D.C.

Grace, N.D. and P.L. Martinson, 1985, The distribution of Mo between the liver and other organs and tissues and live weight gains of grazing sheep, pp. 534–536, C.F. Mills et al. (Eds.), *Trace Elements in Man and Animals* TEMA 5, Commonwealth Agric. Bureaux, Farnham Royal, England.

Gupta, U.C., E.W. Chipman and D.C. Mackay, 1978, Effects of molybdenum and lime on the yield and molybdenum concentration of vegetable crops grown on acid sphagnum peat soil, *Can. J. Plant Sci.* 58: 983–992.

Hambridge, K.M., C. Hambridge, M. Jacobs, and J.D. Baum, 1972, Low levels of zinc in hair, anorexia, poor growth and hypogeusia in children, *Pediatr. Res.* 6: 868–874.

Hambridge, K.M., C.W. Casey and N.F. Krebs, 1987, Zinc, pp. 1–137, W. Mertz (Ed.), *Trace Elements in Human and Animal Nutrition*, Vol. 2, Academic Press, New York.

Hansen, J.C., H.C. Wulf, N. Kormann, and K. Alboge, 1985, Cadmium concentration in blood samples from an East Greenlandic population, *Danish Med. Bull.* 32: 277–279.

Harrison, R.M. and M.B. Chirgawi, 1989, The assessment of air and soil contributors of some trace metals to vegetable plants: II. Translocation of atmospheric and laboratory generated cadmium aerosols to and within vegetable plots, *Sci. Total Environ.* 83: 35–45.

Harter, R.D., 1991, Micronutrient adsorption–desorption reactions in soils, pp. 59–87, in J.J. Mortvedt, F.R. Cox, L.M. Shuman and R.M. Welch (Eds.), *Micronutrients in Agriculture*, Soil Science Society of America, Madison, WI.

Haygarth, P.M., A.F. Harrison, and K.C. Jones, 1995, Plant selenium from soil and the atmosphere, *J. Environ. Qual.* 24: 768–771.

Holmgren, G.G.S., M.W. Meyer, R.L. Chaney, and R.B. Daniels, 1993, Cadmium, lead, zinc, copper and nickel in agricultural soils of the United States of America, *J. Environ. Qual.* 22: 335–348.

Hovmand, M.F., J.C. Tjell and H. Mosbaek, 1983, Plant uptake of airborne cadmium, *Environ. Pollut.* Ser. A. 30: 27–38.

Jacobs, L.W., D.R. Keeney, and L.M. Walsh, 1970, Arsenic residue toxicity to vegetable crops grown on plainfield sand, *Agron. J.* 62: 588–591.

Johnson, W.R. and Proctor, J, 1977, A comparative study of metal levels in plants from two contrasting lead–mine sites, *Plant Soil* 46: 251–257.

Jones, J.B., 1991, Plant tissue analysis in micronutrients, pp. 477–521, J.J. Mortvedt, F.R. Cox, L.M. Shuman and R.M. Welch (Eds.), *Micronutrients in Agriculture*, Soil Science Society of America, Madison, WI.

Kagi, J.H.R. and B.L. Vallee, 1960, Metallothionein: A cadmium and zinc–containing protein from equine renal cortex, *J. Biol. Chem.* 235: 3460.

Kolonel, L.N., 1976, Association of cadmium with renal cancer, *Cancer*, 37: 1782–1787.

Krauskopf, K.B., 1967, *Introduction to Geochemistry,* McGraw–Hill, New York.

Kubota, J., 1977, Molybdenum status of United States soils and plants, W.R. Chappel and K.K. Peterson (Eds.), *Molybdenum in the Environment*, Vol. 2, Marcel Dekker, New York.

Kubota, J., V.A. Lazar, L.N. Langan and K.C. Beeson, 1961, The relationship of soils to molybdenum toxicity in cattle in Nevada, *Soil Sci. Soc. Am. Proc.* 25: 227–232.

Kubota, J., V.A. Lazar, G.H. Simonson and W.W. Hill, 1967, The relationship of soils to molybdenum toxicity in grazing animals in Oregon, *Soil Sci. Soc. Am. Proc.* 31: 667–671.

Lee, K.W. and D.R. Keeney, 1975, Cadmium and zinc additions to Wisconsin soils by commercial fertilizers and wastewater sludge application, *Water, Air, Soil Pollut.* 5: 109–112.

Levender, O.A., 1979, Lead toxicity and nutritional deficiencies, *Environ. Health Perspect.* 29: 115–125.

Levy, L.S., F.J.C. Roe, D. Malcom, G. Kazanatzis, J. Clark, and H.S. Platt, 1973, Absence of prostatic changes in rats exposed to cadmium, *Ann. Occup. Hyg.* 16: 111–118.

Logan, T.J. and R.L. Chaney, 1983, Utilization of municipal wastewater and sludges on land–metals, Workshop on Utilization of Municipal Wastewater and Sludge on Land, Denver, CO.

Logan, T.J., A.C. Chang, A.L. Page and T.J. Ganje, 1987, Accumulation of selenium in crops grown on sludge–treated soil, *J. Environ. Qual.* 16(4): 349–352.

Lucas, P.A., A.G. Jariwalla, J.H. Jones, J. Gough and P.T. Vole, 1980, Fatal cadmium fume inhalation, *Lancet* 2: 205.

Lue–Hing, C., R.I. Pietz, P. Tata, R. Johnson, R. Sustich and T.C. Granato, 1999, A national sewage sludge survey: Quality status relative to the part 503 Rule, WEFTEC '99, New Orleans, Louisiana, Water Environment Federation, Arlington, VA.

Lund, L.J., E.E. Betty, A.L. Page and R.A. Elliott, 1981, Occurrence of naturally high cadmium levels in soils and its accumulation by vegetation, *J. Environ. Qual.* 10: 551–556.

Ma, L.Q., F. Tan and W.G. Harris, 1997, Concentration and distribution of eleven metals in Florida soils, *J. Environ. Qual.* 26: 769–775.

Mahaffey, K.R. and J.E. Vanderveen, 1979, Nutrient–toxicant interactions: Susceptible populations, *Environ. Health Perspect.* 29: 81–87.

Malcom, D., 1972, Potential carcinogenic effect of cadmium in animals and man, *Ann. Occup. Hyg.* 15: 33–37.

Marston, H.R., 1950, Problems associated with copper deficiency in ruminants, W.D. McElroy and B. Glass (Eds.), *Copper Metabolism Symposium on Animal, Plant and Soil Relationships*, Johns Hopkins University Press, Baltimore, MD.

Mayland, H.F., L.F. James, K E. Panter and J.L. Sonderegger, 1989, Selenium in seleniferous environments, 15–50. L.W. Jacobs (Ed.), *Selenium in Agriculture and the Environment*, American Society of Agronomy, Soil Science Society of America, Madison, WI.

McBride, M.B., 1995, Toxic metal accumulation from agricultural use of sludge: Are USEPA regulations protective?, *J. Environ. Qual.* 24: 5–18.

McKenzie–Parnell, J.M., and G. Eyon, 1987, Effect of New Zealand adults consuming large amounts of cadmium in oysters, pp. 420–430, D.D. Hemphill (Ed.), *Trace Substances in Environmental Health–XXI, Proc. of the University of Missouri's 21st Annual Conference on Trace Substances in Environmental Health,* St. Louis, MO.

McKenzie–Parnell, J.M., T.E. Kjellstrom, R.P. Shaarma and M.F. Robinson, 1988, Unusually high intake and fecal output of cadmium and fecal output of other trace elements in New Zealand adults consuming dredge oysters, *Environ. Res.* 46: 1–14.

McLaughlin, M.J., L.T. Palmer, K.G. Tiller, T.A. Beech and M.K. Smart, 1994, Increased soil salinity causes elevated cadmium concentrations in field–grown potato tubers, *J. Environ. Qual.* 23: 1013–1018.

McNeal, J.M. and L.S. Balistrieri, 1989, Geochemistry and occurrence of selenium: An overview, pp. 1–13, in L.W. Jacobs (Ed.), *Selenium in Agriculture and the Environment,* American Society of Agronomy, Madison, WI.

Mermut, A.R., J.C. Jain, L. Sog, R. Kerrich, L. Kozak and S. Jana, 1996, Trace element concentrations of selected soils and fertilizers in Saskatchewan, *J. Environ. Qual.* 25: 845–853.

Mertz, W., 1967, The role of chromium in glucose metabolism, D.D. Hemphill (Ed.), *Trace Substances in Environmental Health — I. University of Missouri Annual Conf., 1st Proc.,* University of Missouri.

Miller, E.R., L. Xingen, and D.E. Ullrey, 1991, Trace elements in animal nutrition, pp. 593–662, J.J. Mortvedt, F.R. Cox, L.M. Schuman and R.M. Welch (Eds.), *Micronutrients in Agriculture,* Soil Science Society of America, Madison, WI.

Millner, J.E., J.J. Hassett, and D.E. Koeppe, 1976, Uptake of cadmium by soybeans as influenced by soil cation exchange capacity, pH and available phosphorus, *J. Environ. Qual.* 5: 157–160.

Mills, C.F. and G.K. Davis, 1987, Molybdenum, pp. 429–463, W. Mertz (Ed.), *Trace Elements in Human and Animal Nutrition,* Vol.1, 5th ed., Academic Press, San Diego, CA.

Moraghan, J.T. and H.J. Mascagni, Jr., 1991, Environmental and soil factors affecting micronutrient deficiencies and toxicities, *Micronutrients in Agriculture,* Soil Science Society of America, Madison, WI.

Mortveldt, J.J., D.A. Mays and G. Osborn, 1981, Uptake by wheat of cadmium and other heavy metal contaminants in phosphate fertilizers, *J. Environ. Qual.* 10: 193–197.

Mortvedt, J.J., F.R. Cox, L.M. Shuman, and R.M. Welch, 1991, *Micronutrients in Agriculture,* Soil Science Society of America, Madison, WI.

Moss, L.H., E. Epstein and T. Logan, 2002, Comparing the characteristics, risk and benefits of soil, amendments and fertilizers used in agriculture. 16th Annual Residuals and Biosolids Management Conference — Privatization, Innovation and Optimization: How to Do More for Less, Austin, Texas, Water Environment Federation, Alexandria, VA.

NRC, 1974, Chromium, National Research Council, 155. National Academy of Sciences, Washington, D.C.

NRC, 1980, Mineral Tolerance of Domestic Animals, National Research Council, National Academy of Sciences, Washington, D.C.

Nielsen, F.H., 1984, Ultratrace elements in nutrition, *Annu. Rev. Nutr.* 4: 21–41.

Olsen, S.R. and F.W. Watanabe, 1979, Interaction of added gypsum in alkaline soils with uptake of iron, molybdenum, manganese and zinc by sorghum, *Soil Sci. Soc. Am. J.* 43: 125–130.

Onken, B.M. and L.R. Hossner, 1995, Plant uptake and determination of arsenic species in soil solution under flooded conditions, *J. Environ. Qual.* 24: 373–381.

Page, A.L., 1974, Fate and effects of trace elements in sewage sludge when applied to agricultural lands, Report No. EPA–670/2–74–005. U.S. Environmental Protection Agency, Washington, D.C.

Page, A.L., F.T. Bingham, and A.C. Shang, 1981, Cadmium, pp. 77–109, N.W. Lepp (Ed.), *Effect of Heavy Metal Pollution on Plants*, Vol. 1. Applied Science, Barking, Essex, England.

Pais, I. and J. Jones, J.B., 1997, *The Handbook of Trace Elements*, St. Lucie Press, Boca Raton, FL.

Parr, J.F., E. Epstein, R.L. Chaney and G.B. Willson, 1977, Impact of the disposal of heavy metals in residues on land and crops, pp. 126–133, *Proc. 1977 National Conference on Treatment and Disposal of Industrial Wastewaters and Residues, Houston, Texas.* Information Transfer, Rockville, MD.

Payne, C.G., 1977, Involvement of molybdenum in feather growth, *Br. Poult. Sci.* 18: 427–432.

Pierce, F.J., H. Dowdy and D.F. Grigal, 1982, Concentrations of six trace metals in some major Minnesota soil series, *J. Environ. Qual.* 22: 805–811.

Prasad, A.D., A. Miale, Jr., A. Farid, H.H. Sandstead and R. Schulert, 1963, Zinc metabolism in patients with the syndrome of iron deficiency anemia, hepatosplenomegaly, dwarfism and hypogonadism, *J. Lab. Clin. Med.* 61: 537–549.

Pulido, P., J.H.R. Kagi, and B.L. Vallee, 1966, Isolation and some properties of human metallothionein, *Biochemistry* 5: 1768.

Ryan, J.A., H.R. Pahren, and J.B. Lucas, 1982, Controlling cadmium in the human food chain: A review and rationale based on health effects, *Environ. Res.* 28: 251–302.

Reddy, J., D. Svoboda, D. Azarnoff, and R. Dowas, 1973, Cadmium induced Leydig cell tumors of rat testis: Morphologic and cytochemical study, *J. Nat. Cancer Inst.* 51: 891–903.

Römheld, V. and H. Marschner, 1991, Function of micronutrients in plants, 297–328, J.J. Mortvedt, F.R. Cox, L.M. Shuman and R.M. Welch (Eds.), *Micronutrients in Agriculture*, 2nd ed., Soil Science Society of America, Madison, WI.

Scheinberg, H.I., 1969, The essentiality and toxicity of copper in man, *Trace Substances in Environmental Health* – III. *Annu. Conf. 3rd. Proc.* University of Missouri, Columbia.

Schnitzer, M., 1978, Humic substances, M. Schnitzer and S.U. Kaha (Eds.), *Soil Organic Matter,* Elsevier, New York.

Selby, L.A., A.A. Case, C.R. Dorn, and D.J. Wagstaff, 1974, Public health hazards associated with arsenic poisoning in cattle, *J. Am. Vet. Med. Assoc.* 165: 1010–1014.

Shacklette, H.T., 1960, Soil and plant sampling at the Mahoney Creek lead–zinc deposit, Revillagigedo Island, Southeastern Alaska, U.S. Geological Survey Prof. Paper 400–B, U.S. Geological Survey, Washington, D.C.

Shacklette, H.T., 1970, Mercury content of plants, U.S. Geological Survey, Washington, D.C.

Shaikh, Z.A. and J.C. Smith, 1980, Metabolism of orally ingested cadmium in humans, p. 569, B. Holmstedt, R. Lauwerys, M. Mercer and M. Rocker (Eds.), *Mechanism of Toxicity and Hazard Evaluation,* Elsevier, Amsterdam.

Sharma, R.P., T. Kjellstrom and J.M. McKenzie, 1983, Cadmium in blood and urine among smokers and nonsmokers with high cadmium uptake via food, *Toxicology* 29: 163–171.

Sherlock, J.C., 1984, Cadmium in foods and diet, *Experientia.* 40: 152–156.

Shuman, L.M., 1991, Chemical forms of micronutrients in soil, pp. 113–144, J.J. Mortvedt, F.R. Cox, L.M. Shuman and R.M. Welch (Eds.), *Micronutrients in Agriculture*, 2nd ed., Soil Science Society of America, Madison, WI.

Staker, E.V. and R.W. Cummings, 1941, The influence of zinc on the productivity of certain New York peat soils, *Soil Sci. Soc. Am. Proc.* 6: 207–214.

Stevens, D.R., L.M. Walsh, and D.R. Keeney, 1972, Arsenic phytotoxicity on Plainfield soil as affected by ferric sulfate or aluminum sulfate, *J. Environ. Qual.* 1: 301–303.

Sunderman, F.W.J., 1971, Metal carcinogenesis in experimental animals, *Food Cosmet. Toxicol.* 9: 105–120.

Sunderman, F.W.J., 1978, Carcinogenic effect of metals, *Fed. Proc.* 37: 40–46.

Swaine, D.J. ,1955, The trace element content of soils, *Bur. Soil Sci.* 48: 1–157.

Toepfer, E.W., W. Mertz, M.M. Polansky, E.E. Roginski and W.R. Wolf, 1977, Preparation of chromium–containing material of glucose tolerance factor activity from brewer's yeast extracts and by synthesis, *J. Agr. Food Chem.* 25: 162–166.

Tsutsumi, M., 1981, Arsenic pollution of arable land, pp. 181–182, K. Kitagishi and I. Yamane (Eds.), *Heavy Metal Pollution in Soils of Japan*, Japan Soil Science Press, Tokyo,

Underwood, E.J., 1977, *Trace Elements in Human and Animal Nutrition*, 4th ed., Academic Press, New York.

Underwood, E.J., 1981, *The Mineral Nutrition of Livestock*, 2nd ed., Commonwealth Agriculture Bureaux, Slough, England.

U.S. Department of Agriculture, 1968, Summary of registered agricultural pesticide chemical uses, U.S. Department of Agriculture, Pesticide Regulation Division, Looseleaf N.P. Washington, D.C.

USEPA, 1990, National sewage sludge survey: Availability of information and data and anticipated impacts on proposed regulations: Proposed Rule 40 CFR Part 503, *Fed. Reg.* 55: 47210–47283.

Van Campen, D.R., 1991, Trace elements in human nutrition, pp. 663–701, J.J. Mortvedt, F.R. Cox, L.M. Shuman and R.M. Welch (Eds.), *Micronutrients in Agriculture*, Soil Science Society of America, Madison, WI.

Water Environment Association of Ontario, 2001, Fate and significance of selected contaminants in sewage biosolids applied to agricultural land through literature review and consultation with stakeholder groups, Water Environment Association of Ontario, Toronto.

Webber, M.D. and J. Nichols, 1995, Organic and metal contaminants in Canadian municipal sludges and sludge compost, Wastewater Technology Centre, Burlington, Ontario.

Welch, R.M., W.H. Allaway, W.A. House and Kubota, 1991, Geographic distribution of trace element problems, pp. 31–57, J.J. Mortvedt, F.R. Cox, L.M. Shuman and R.M. Welch (Eds.), *Micronutrients in Agriculture*, Soil Science Society of America, Madison, WI.

Wells, B.R. and J.I. Gilmore, 1977, Sterility in rice cultivars as influenced by MSMA rate and water management, *Agron. J.* 69: 451–454.

WHO, 1982, Assessment of human exposure to cadmium and lead through biological monitoring, World Health Organization. Report EFP/82–26, Geneva, Switzerland.

WHO, 1990, Environmental health criteria 101: Methylmercury, World Health Organization, Geneva, Switzerland.

Woolson, E.A., 1983, Emissions, cycling and effects of arsenic in soil ecosystems, B.A. Fowler (Ed.), *Biological and Environmental Effects of Arsenic*, Elsevier Science, New York.

Woolson, E.A., J.H. Axley and P.C. Kearney, 1971, The chemistry and phytotoxicity of arsenic in soils: I. Contaminated field soils, *Soil Sci. Soc. Am. Proc.* 35: 938–943.

Yamagata, N. and I. Shigematsu, 1970, Cadmium pollution in perspective, *Bull. Inst. Public Health* 19: 1–27.

Yang, S.P., M.M. Luo and H.J. Wei, 1985, Effect of molybdenum and tungsten on carcinogenesis, 160–161, in C.F. Mills et al. (Eds.), *Trace Elements in Man and Animals, TEMA 5*, Commonwealth Agriculture Bureaux, Farnham Royal, England.

CHAPTER 5

The Effect of Sewage Sludge and Biosolids on Uptake of Trace Elements and Reactions in Soil

INTRODUCTION

The reactions of trace elements and plant micronutrients in soil and their uptake and accumulation by plants affect the availability to animals and humans. The state of trace elements in soils generally falls into the following categories:

- Soluble
- Exchangeable
- Insoluble inorganic precipitates
- Chelated/complexed on organic matter or its fractions
- Adsorbed on inorganic colloids
- As minerals

The uptake of trace elements and plant micronutrients depends on the soluble equilibrium and exchangeability of an element and, therefore, its concentration in the soil solution. Adsorption mechanisms are important in determining the availability of the metal to plants. In addition to adsorption by soil organic matter, metal adsorption by reactive soil surfaces is important. Corey et al. (1981) indicated that adsorption on mineral surfaces is probably the dominant process controlling metal solution activities. Activity is generally considered as effective concentration.

When metals are added through the application of biosolids, solution metal activities would be controlled by adsorption to biosolids and soil mineral surfaces, as long as the amount of metal added did not exceed the capacity of specific surface adsorption sites in the soil. Some have suggested that the metal chemistry of a particular biosolid may be important in the availability of the metal to plants.

Heavy metals can also revert with time to chemical forms less available to plants (Epstein and Chaney, 1978). Sposito et al. (1983) found that Zn from biosolids reverted to the less available carbonate form over time. However, metal reversion in soil is small and/or slow and can be reversed by soil acidity (Logan and Chaney, 1983). The form of the heavy metal in soil affects its solubility and, therefore, the potential for movement through the soil and uptake by plants. This section discusses the regulated heavy metals and other trace elements in soils and plants. Throughout this chapter, the term "trace elements" will be used except when a specific literature citation uses the term "heavy metals."

The soil factors that affect trace element uptake are:

- Soil pH — Soil pH is the most important factor controlling metal solubility. The concentration in the soil solution is governed by an equilibrium between adsorbed and dissolved forms. With the exception of Mo and Se, all the essential trace elements are more soluble at low pH. The solubility of metal carbonates, phosphates and sulfides is increased at low pH (Lindsey, 1979). Heavy metals are more available to plants below pH 6.5 (Chaney, 1973; CAST, 1980). The solubility of most heavy metals or trace elements increases as the acidity of the soil increases. There is an approximately 100-fold decrease in activity (effective concentration) of Zn and Cu for each unit increase in soil pH. At low pH, heavy metals are more soluble (in the soil solution) and mobile. This increases their potential uptake by plants and potential for movement through the soil profile. Anderson and Christensen (1988) showed the pH is more important than any other single property in controlling Zn mobility.
- Organic matter — Organic matter can bind and complex certain heavy metals so that they are less available to plants. Organic matter has a high cation exchange capacity (CEC) compared to the mineral fraction of soils. This factor is important in binding such heavy metals as Cu, Zn, Ni and Cd. Organic matter can also chelate heavy metals, reducing their availability to plants. This property might be more important than the simple cation-exchange role of the organic matter.
- Elemental interactions — Certain elements can interact with some heavy metals to decrease their availability to plants. The availability of heavy metals to plants and their mobility through the soil is dependent on interactions with other elements. The hydrous oxides of Fe and Mn can control the availability of heavy metals by sorption and desorption (Jenne, 1968; Quirk and Posner, 1975). Phosphorus combines with metal ions to form soluble or insoluble complexes (Epstein and Chaney, 1978). Barrow (1987) showed that orthophoshates could either increase or decrease Zn retention in soil depending on pH. McBride (1985) reported that there was a decrease in sorption of Cu in soils when orthophosphates were present. Cadmium retention in the soil is influenced by Fe content.
- Soil Cation Exchange Capacity (CEC) — This factor is important in the binding of heavy metals and other cations (Chaney, 1973; CAST, 1976). This includes all the regulated trace elements except those that occur as anions in the soil solution. Soils with low CEC such as sands have a much lower binding power as compared to clays with a high CEC. Organic matter has a high CEC and contributes to the total CEC of soils and the binding power to heavy metals. Recognition of the importance of CEC in the availability of metals to plants resulted in early recommendations and regulations relating heavy metal additions to land based on CEC (USEPA, 1980). Three CEC categories of soil were established: <5, 5–15, >15 meq/100 g soil. However, at a later date it was difficult to establish a direct

relationship between CEC and metal uptake. Consequently, in the 40 CFR 503 regulations, this relationship was omitted.
• Soil water, soil temperature and soil aeration — These factors affect heavy metal uptake by plants. Soil water affects the amount of trace elements in the soil solution. As the soil dries, the heavy metals in solution become more concentrated and can precipitate or be adsorbed on soil colloids and become unavailable to plants. Soil water can affect the chemical state of the metal.

Epstein (1971) showed that temperature increased the concentration of Mn, P and Ca in potato tops and roots. Soil aeration affects the oxidation and reduction conditions in the soil. Under submerged conditions, reduction occurs with a decrease in Eh and an increase in pH (Lindsay, 1979). Under reduced conditions, probably only Fe and Mn are more apt to be available to plants.

PLANT UPTAKE OF HEAVY METALS

Plant uptake of trace elements affects the concentration of the metal in plant tissues and can affect human and animal nutrition or phytotoxicity. The uptake of trace elements by plants involves several mechanisms. For uptake by plant roots, the trace element must move from the soil to the plant root. The principal mechanisms are convection and diffusion (Chaney, 1975). As water moves through the soil, it solubilizes plant nutrients and trace elements, including heavy metals. Plants absorb water and through convective flow and metals enter the root system. At the root surface elemental movement is governed by the rate of diffusion. One of the more important phenomena that affect trace element movement to roots is chelation. Organic matter is the most important chelating agent. Once the trace element enters the root system, at least two phases involve movement and accumulation in upper plant tissues (Chaney, 1975). The first phase involves movement to and release into the xylem sap and the second involves movement in the xylem sap to plant tissues. Uptake of water and ions occurs in two separate pathways (Kochian, 1991). A detailed discussion of micronutrient reactions in soil and uptake by plants can be found in *Micronutrients in Agriculture* (Mortvedt et al., 1991).

Chaney (1980) introduced the concept of "soil-plant-barrier." Chaney (1983) states a soil-plant-barrier protects the food chain from toxicity of a trace element when one or more of the following processes limits maximum levels of that element in edible plant tissues to levels safe for animals:

• Insolubility of the element in soil prevents uptake
• Immobility of an element in fibrous roots prevents translocation to edible plant tissues
• Phytotoxicity of the element occurs at concentrations where the element, in edible plant tissues, is below a level that is injurious to animals.

Chaney (1994) states that the soil-plant-barrier concept showed that soil and plant chemistry prevents risk to animals from nearly all biosolids-applied trace elements mixed in the soil.

Figure 5.1   Effect of nickel toxicity on snap beans. Dosage of nickel ranges from 0 to 235.0 mg/kg. (Courtesy of Dr. R. Chaney, USDA.)

Phytotoxicity can be manifested by excessive trace element adsorption and result in injury or death to plants. In biosolids, the elements most likely to be phytotoxic are Cu, Ni and Zn. Cu and Ni toxicity retards growth and can inhibit Fe translocation, resulting in Fe deficiency and chlorosis. The effect of nickel on phytotoxicity of snap beans is shown in Figure 5.1. Zn toxicity can also result in retarded growth and symptoms similar to Fe deficiency. Phytotoxicity is affected by soil pH, crop species and cultivars, biosolids' metal concentration and other soil and climatic factors. Table 5.1 shows the relative sensitivity of crops to biosolid-applied trace elements.

The uptake of trace elements by plants is a function of the concentration of the element, soil characteristics, climate, plant species, or cultivars (Chang et al., 1982, 1987; Giordano et al., 1979; Hinsely et al., 1982). Chaudri et al. (2001) indicated that uptake of Cd by wheat was strongly influenced by soil pore water. Bingham et al. (1975) showed that Cd content of plants varies according to plant species and plant tissue as shown in Figure 5.2. Cereals and legumes accumulated less Cd in the shoot than leafy vegetables.

Kim et al. (1988) evaluated the relative concentrations of Cd and Zn in tissues of 12 selected food plants. They found that the relative concentration of these metals in several sludge-treated soils was significantly different. Figure 5.3 shows the cadmium content of harvested plant parts as related to soil cadmium. Higher Cd in the soil resulted in a higher Cd content in plant organs. The leaves contained the highest Cd content, followed by the root or tuber, with the least in the fruit, seed,

Table 5.1 Relative Sensitivity of Crops to Biosolids-Applied Trace Elements*

Very sensitive[1]	Sensitive[2]	Tolerant[3]	Very tolerant[4]
Chard	Mustard	Cauliflower	Corn
Lettuce	Kale	Cucumber	Sudan grass
Red beet	Spinach	Zucchini squash	Smooth bromegrass
Carrot	Broccoli	Flat pea	Red fescue
Turnip	Radish	Oat	
Peanut	Tomato	Orchard grass	
Lidino clover	Zigzag, Red Kura and crimson clover	Japanese bromegrass	
Alsike clover	Alfalfa	Switchgrass	
Crown vetch	Korean lespedeza	Redtop	
Alfalfa	Sericea lespedeza	Buffalo grass	
White sweet clover	Blue lupine	Tall fescue	
Yellow sweet clover	Bird's-foot trefoil	Red fescue	
Weeping love grass	Hairy vetch	Kentucky bluegrass	
Lehman love grass	Soybean		
Deertongue	Marigold		
	Snapbean		
	Timothy		
	Colonial bent grass		
	Perennial ryegrass		
	Creeping bent grass		

* Sassafras sandy loam amended with highly stabilized and leached digested biosolids containing 5300 mg Zn, 2400 mg Cu, 320 mg Ni, 390 mg Mn, 23 mg Cd/kg dry biosolids.
[1] Injured at 10% of a high metal biosolid at pH 6.5 and pH 5.5.
[2] Injured at 10% of a high metal biosolid at pH 5.5 but not at pH 6.5.
[3] Injured at 25% high metal biosolid at pH 5.5, but not at pH 6.5; and not at 10% biosolid at pH 5.5 or 6.5.
[4] Not injured even at 25% biosolid, pH 5.5.
Source: Courtesy of Dr. R. Chaney, USDA.

or flower. Table 5.2 shows the relative concentrations of Cd in various plants with respect to the Cd content of the soil.

Jing and Logan (1992) showed that the chemistry of the biosolids and its trace element concentration had some effect on plant uptake. They found that Cd concentrations in Sudax [*Sorghum bicolor* (L.) Moench] taken up from 17 anaerobically digested biosolids were highly correlated with total and resin-extractable biosolid Cd, even though the total Cd application to the soil was constant.

Table 5.3 shows the concentration of cadmium, zinc, copper and nickel in edible portions of vegetables grown on sewage sludge from 1977 through 1979 (Corey et al., 1987). The data illustrate several important aspects of metal uptake.

- Leafy vegetables such as Swiss chard and spinach accumulated more Cd and Zn than non-leafy vegetables.
- The uptake of Cu and Ni was not much different in all the vegetables. Thus, crop uptake varies from one metal to another.
- Root crops such as beets and carrots did not accumulate more metals than above-ground edible organs such as tomatoes and green beans.

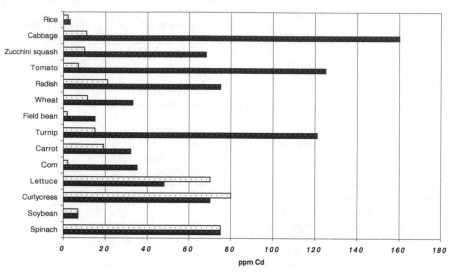

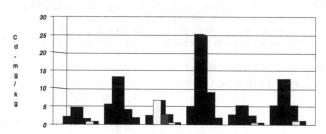

Figure 5.2 Cadmium content in edible and nonedible portions of plants. (Adapted from Bingham et al., 1975.)

	Las Virgines I	Las Virgines II	Greenfield I	Greenfield II	Domino I	Domino II
■ Soil	2.1	5.6	2.4	5.1	2.8	5.3
■ Leaves	4.69	13.23	6.71	25.08	5.43	12.7
■ Root/Tuber	1.73	4.14	2.73	9.12	2.49	5.4
Fruit/Seed/Flower	0.89	1.94	0.44	2	0.5	1.1

Figure 5.3 Cadmium content of harvested plants as related to soil cadmium. (Adapted from Kim et al., 1988.)

　　Brown et al. (1996) evaluated the uptake of Cd by a variety of fruits and vegetables grown on long-term biosolids-amended soils. They found that it was possible to establish a useful, quantitative relationship between the concentrations of Cd in lettuce to the other vegetables in the study. They suggested that this relationship could establish a means of assessing possible risk associated with Cd contamination.

　　Relatively few long-term biosolids field studies took place during the 1970s and 1980s. Page et al. (1987) indicated that several long-term studies on plant uptake of metals from sewage sludge/biosolids revealed some important findings:

Table 5.2   Relative Concentrations of Cd in Plant
Organs for the Low- and High-Cd Las
Virgenes Soils, Using Swiss Chard as
the Reference Plant

Plant	Las Virgenes I 2.1 mg/kg Cd	Las Virgenes II 5.6 mg/kg Cd
Swiss chard	1.0	1.0
Tomato leaf	2.5	0.2
Tomato fruit	0.2	0.01
Pepper leaf	8.5	3.3
Pepper fruit	1.6	0.7
Leaf lettuce	3.1	2.7
Head lettuce	2.8	3.3
Radish leaf	2.0	2.4
Radish tuber	0.6	0.6
Potato leaf	0.54	0.79
Potato tuber	0.41	0.55
Corn leaf	1.0	0.5
Corn kernel	0.1	0.06
Wheat leaf	0.6	0.2
Wheat grain	0.2	0.1
Broccoli leaf	0.6	0.9
Broccoli flower	0.1	0.4
Carrot leaf	3.0	1.3
Carrot root	1.8	0.9
Beet leaf	1.5	1.4
Beet root	0.5	0.2

Source. Adapted from Kim et al., 1988.

- Plant uptake is a function of plant species, individual trace elements, soil characteristics and sludge/biosolids characteristics.
- Sludge/biosolids is both a source and a sink for trace elements.
- Trace element uptake by plants may obey many different rate response functions: linear, sympatric, no response, or even negative.

Subsequently, many researchers have found that the uptake by various crops was not linear with trace elements or sludge/biosolids' application rate, but rather approached a maximum and then leveled or decreased (Chaney and Ryan, 1992; Logan and Chaney, 1983; Corey et al., 1987). This phenomenon was called the plateau response. For low-metal biosolids, the phytoavailability is controlled by the biosolids' chemistry (Brown et al., 1998).

Beckett et al. (1979) suggested that the organic matter in biosolids is responsible for the binding effect of metals. McBride (1995) contended that as the organic matter decomposed, trace elements will be released into more soluble forms and result in increased uptake from biosolids. He termed this hypothesis as the "time bomb" effect. Chang et al. (1997) indicated that the "sludge time bomb hypothesis" may show a plateau effect during the course of biosolids' application. The issue is what happens when the addition of biosolids organic matter decomposes. Once the biosolids organic matter is lost, will the metals become more available? McBride (1995) stated, "Because soils have a finite capacity to immobilize metal by adsorption or

Table 5.3 Concentration of Cadmium, Zinc, Copper and Nickel in Edible Portions of Plants Grown on Sludge-Amended Soil

Crop	Amount of Biosolids Applied to Land – tonnes/ha				
	0	60	120	240	300
	Cadmium – mg/kg in edible tissue				
Beets	0.2	1.4	1.6	2.7	2.9
Tomatoes	1.1	1.8	2.4	2.2	3.4
Swiss chard	2.3	8	12.2	16.8	22.1
Carrots	0.7	1.4	1.2	1.9	2.3
Green beans	0.4	0.3	0.4	0.4	0.5
Spinach	6.4	12.6	10.3	14.4	12.1
	Zinc – mg/kg in Edible Tissue				
Beets	34	45	62	93	90
Tomatoes	27	30	30	34	35
Swiss chard	69	129	176	237	302
Carrots	20	22	25	27	32
Green beans	32	33	37	37	37
Spinach	147	249	265	309	311
	Copper – mg/kg in Edible Tissue				
Beets	8.8	10.4	9.8	11.1	12.8
Tomatoes	13.2	16.6	14.5	16.2	16.6
Swiss chard	23.7	20.8	25.4	30.9	29.2
Carrots	5.4	6.0	5.8	5.9	6.5
Green beans	8.6	8.9	9.2	8.2	8.5
Spinach	13.8	14.2	17.3	18.9	19.1
	Nickel – mg/kg in Edible Tissue				
Beets	0.5	0.6	0.9	1.4	1.5
Tomatoes	1.1	1.3	1.4	4.1	2.7
Swiss chard	1.3	1.4	2.5	3.5	4.3
Carrots	1.9	0.7	0.8	1.2	1.0
Green beans	3.3	1.2	2.4	3.6	3.3
Spinach	1.4	1.5	1.7	2.7	2.6

Source: Corey et al., 1987, pp. 25–51, A.L. Page et al. (Eds.), *Land Application of Sludge*, Lewis Publishers, Chelsea MI. With permission.

precipitation reactions, without the protective effect of the sorptive material in the sludge itself, a Langmuir-type relationship would be expected."

The plateau effect implies that biosolids are both a source and a sink for trace elements applied to soil. At low biosolids' application, the soil binds the elements and plant uptake is linear. At very high biosolids' application rates, the biosolid matrix influences the binding of trace elements. The plateau theory, therefore, posits that the concentrations of heavy metals in plant tissue will reach a plateau as biosolids mass loadings are increased and they will remain at a plateau after the termination of biosolids' application. Chang et al. (1997) used a set of experimental data obtained from a 10-year field biosolids' land application to evaluate the hypotheses of the plateau and the time bomb. They indicated that with those set of data, an actual plateau or time bomb was not evident. Biosolids' application had reached 2880

Mg/ha, which probably represented a worst-case scenario in terms of pollutant loading.

However, there are other factors that affect the binding of trace elements in the soil. Furthermore, the rate of soil organic matter decomposition will be rapid initially when the soluble carbohydrates, fats and some proteins are microbially metabolized. Once these fractions are assimilated by microorganisms, then hemicellulose, cellulose and lignin — that are very significant in metal binding — are slow to decompose.

Complexation of trace elements is with the humified fraction in the soil that decomposes very slowly. The organic fraction in biosolids is generally 50% to 60%. The remainder for the most part is inert. Approximately 40% to 50% of this organic matter is fairly rapidly decomposed. The remainder is very slow to decompose. Therefore, it is highly unlikely that all the organic matter will disappear and make the bound trace elements available for plant uptake.

Several studies have shown that organic matter from sludge and manure additions to soils may still be present in significant quantities in the soil for long periods of time. As much as 50% of organic carbon remained in the soil after 10 years of land application sludge or manure (Johnston, 1975; Johnston and Wedderburn, 1975).

Brown et al. (1998) found that, 16 years following application, there was no significant increase in Cd by lettuce, even though a significant portion of the organic carbon from the biosolids had disappeared. They indicated that the addition of biosolids altered the soil chemistry so that the incremental increase in plant availability of soil Cd in the soil/biosolids mixture was less than that of the soil alone. They further determined that increases in Cd adsorption in biosolids-amended soils appeared to be related to the inorganic complexing ability added to the soil with biosolids.

Sloan et al. (1998) evaluated the recovery of biosolids-applied heavy metals 16 years after application. The results of the study showed that biosolids organic matter decomposes slowly when applied to a well-drained silt loam in a temperate climate. Therefore, the rapid release of biosolids-derived heavy metals is unlikely to occur.

In an earlier paper, Sloan et al. (1997) reported that biosolids-applied Cd existed in chemical forms that were easily extracted from soil and were readily available for uptake by romaine lettuce. The most easily extracted forms of Cd (i.e., exchangeable and specifically adsorbed) accounted for approximately 75% of total Cd in biosolids-amended soil. Biosolids-applied Cr, Cu, Ni, Pb and Zn were in relatively stable soil chemical forms and were not correlated to plant uptake. Concentrations of Cd, Zn, Cu, Ni and Cr in romaine lettuce were positively correlated to total concentrations of the respective metals in soil.

Data by Bidwell and Dowdy (1987) showed that Cd and Zn availability to corn, following termination of land application of sewage sludge, decreased with time. Corn was sampled for 6 years after termination of three annual applications of sewage sludge. Cumulative sludge applications totaled 0, 60, 120 and 180 Mg/ha.

Table 5.4a Concentration of Cadmium in Corn Stover and Grain for Six Years Following Three Annual Applications of Biosolids Sludge to Soil

Year	Control	Low	Medium	High
		Corn Stover		
1	0.23	3.41	4.96	9.83
2	0.18	1.48	3.22	5.18
3	0.11	0.82	1.72	4.56
4	0.06	1.15	1.78	2.14
5	0.08	1.82	3.35	3.26
6	0.08	0.88	1.26	1.93
		Corn Grain		
1	<0.06	0.10	0.11	0.27
2	<0.04	<0.05	0.07	0.06
3	<0.02	<0.03	0.07	0.15
4	<0.01	<0.03	0.02	0.05
5	<0.02	0.04	0.04	0.06
6	0.02	0.03	0.04	0.06

Source: Adapted from Bidwell and Dowdy, 1987.

Table 5.4b Concentration of Zinc in Corn Stover and Grain for Six Years Following Three Annual Applications of Sewage Sludge to Soil

Year	Control	Low	Medium	High
		Corn Stover		
1	19.9	88.9	101	140
2	31.9	72.8	119	153
3	18.7	66.5	87.3	121
4	30.8	88.0	105	107
5	25.7	112	114	154
6	18.5	66.6	75.8	105
		Corn Grain		
1	25.6	37.0	37.4	44.7
2	32.8	41.0	42.1	50.7
3	27.6	39.3	46.8	72.2
4	35.9	45.2	39.5	43.3
5	36.7	55	55.4	55.3
6	29.6	37.3	38.8	39.4

Source: Adapted from Bidwell and Dowdy, 1987.

Tables 5.4a and 5.4b show some of their data. The data highlight several points:

- Cadmium and Zn concentrations in corn stover and grain increase with sludge applications where high levels of the metals are applied to the soil.
- There was a decrease in the concentration of Cd and Zn in both the corn stover and grain, with time, following application of sewage sludge.
- The corn grain contained significantly lower concentrations of the metals than the stover. In fact, for most of the treatments and time, there was no difference in Cd concentration in the grain.

Logan (1997) conducted a field study from 1991 to 1995. He found that different trace elements behaved differently following biosolids' application. Cadmium, Cu and Zn concentrations in corn increased significantly with biosolids' application, while Ni and Pb levels were lower than the control. Cadmium, Cu and Zn concentrations in corn exhibited a plateau-type response. Lettuce concentrations increased linearly with biosolids' application for Cd, Cu and Zn in all years; linear regression slopes generally declined and stabilized after the first 2 years.

Table 5.5   Mean Concentration of Trace Metals in Corn Stover and Grain Over Time

Year	Treatment	Zn	Cu	Cd	Pb	Ni	Cr
		Stover Concentration - mg/kg					
1–5	Control	16.1	5.0	0.14	1.9	0.7	1.0
6–10		17.7	6.0	0.10	0.9	0.8	1.4
11–19		18.0	8.4	0.16	0.9	0.7	0.9
1–5	Biosolids	23.0	5.8	0.16	1.6	0.8	1.0
6–10		60.2	7.2	0.18	1.0	1.2	2.8
11–19		46.5	7.0	0.18	0.8	0.6	0.4
		Grain Concentration – mg/kg					
1–5	Control	21.9	1.7	0.07	0.6	0.4	0.2
6–10		23.2	2.4	0.03	0.3	0.3	0.2
11–19		20.0	3.2	0.29	0.4	0.4	0.2
1–5	Biosolids	26.9	1.6	0.07	0.7	0.3	0.3
6–10		31.9	1.8	0.03	0.3	0.3	0.2
11–19		26.0	3.2	0.31	0.5	0.3	0.2

Source: Dowdy et al., 1994, pp. 149–155, C.E. Clapp et al. (Eds.), *Sewage Sludge: Land Utilization and the Environment*, American Society of Agronomy, Madison, WI. With permission.

For 20 years, biosolids were applied annually on the Rosemount, Minnesota watershed (Dowdy et al., 1994). The cumulative biosolids and trace metals applied to corn are shown in Table 5.5.The rate of application was based on the N utilization rates for continuous corn production. Corn stover and grain did not have increased concentrations of biosolids-borne Cd, Cr, Cu, Ni, or Pb during 19 years of continuous corn production.

Molybdenum is highly soluble and therefore its potential uptake by plants can be great. This is particularly true at high pH. High soil pH favors low Mo sorption and high plant availability (O'Conner et al., 2001). Because Mo toxicity occurs in cattle and other ruminants, its uptake by forage plants is important. Molybdenum toxicity has not been reported for humans (Ward, 1994). Its excess in the diet can induce Cu deficiency, resulting in a disease to ruminants known as molybdenosis. O'Conner et al. (2001) reported low Mo accumulation by corn stover, even at high biosolids' application and soil loads near 18 kg Mo ha^{-1}.

REACTIONS AND MOVEMENT IN SOILS

In the previous section of this chapter, the chemical forms of trace elements and heavy metals as related to plant uptake were discussed. Another aspect of biosolids-applied trace elements and heavy metals is their potential for movement beyond the root zone.

The nature of the chemical form of the trace element and its association with inorganic and organic soil constituents greatly affects its potential movement, retention and plant uptake (Lake et al., 1984). Soil trace elements are found in a variety of physicochemical forms (Berti and Jacobs, 1996):

- Free or complexed ions in soil solution
- Adsorbed on the surface of clays, Fe and Mn oxyhydroxides, or organic matter that are easily exchangeable
- Present in the lattice of secondary minerals such as phosphates, sulfides, or carbonates
- Occluded in amorphous materials such as Fe and Mn oxyhydroxides, Fe sulfides, or carbonates
- Present in the crystal lattices of primary minerals

To move through the soil or be taken up by plants, a metal has to be in the soil solution. Many of these physicochemical forms complex or bind the metal. There have been numerous attempts made to fractionate and identify the various forms of trace elements. Extraction with nitric acid and diethylenetrinitrilopentaacetic (DTPA) has been used to approximate total metals and plant-available metals in soils, respectively (Sposito et al., 1982; O'Conner, 1988).

An extensive review of the literature on fractionation, characterization and speciation of trace elements and heavy metals in sewage sludge and biosolids and biosolids-amended soil was conducted by Lake et al. (1984). Berti and Jacobs (1996) reported that no fractionation scheme is totally effective in dissolving each distinct form of a trace element. They indicated that the toxicity characteristic leaching procedure (TCLP), used to characterize leaching of heavy metals from wastes deposited in landfills, is not adequate for determining the bioavailability in soils. Baveye et al. (1999) reported that extractions using nitric acid and DTPA remove only a portion of the total metal in soils.

Dowdy and Volk (1983) reviewed the literature to identify conditions where movement and leaching of biosolids-borne heavy metals might occur. They found very little evidence of heavy metal movement beyond the zone of biosolids incorporation. Other studies (Chang et al., 1984; Williams et al., 1987) also reported that there was very little movement of heavy metals into the subsoil or below 30 cm. Dowdy et al. (1991) found that Cd and Zn concentrations were higher in the 0.32 to 0.62 m (1 to 2 ft) depth as compare to the control. Zinc concentrations increased with biosolids' application. This was not true of Cu. Relatively small amounts of biosolids-borne Cd and Zn were found in the subsoil after a 14-year period of massive biosolids' application.

CONCLUSION

Many soil and plant factors affect the bioavailability of trace elements and heavy metals to plants. Plant uptake is a function of plant species, individual trace elements, soil characteristics and sludge/biosolids characteristics. Accumulation in plants varies with the species, cultivars and organs within the plant. In general, more trace elements, including heavy metal, is concentrated in the leaves than in the fruit or grain.

A major controversy focuses on what happens to heavy metals with time. In a theory termed the "time bomb" effect, some scientists hypothesized that as organic matter decomposes, trace elements would be released into more soluble forms and result in increased uptake by plants from biosolids. Currently, long-term studies do not show that this is occurring.

Many researchers have found that the uptake of trace elements and regulated heavy metals by various crops was not linear with trace elements or sludge/biosolids' application rate; rather it approached a maximum and then leveled or decreased. This phenomenon is called the "plateau response."

Trace elements from land-applied biosolids do not appear to move through the soil profile and essentially remain in the top 30 cm. This greatly reduces the potential for heavy metal to move through the soil into ground water.

REFERENCES

Anderson, P.R. and T.H. Christensen, 1988, Distribution coefficients of Cd, Co, Ni and Zn in soils, *J. Soil Sci.* 39: 15–22.

Barrow, N.J., 1987, The effects of phosphate on zinc sorption by a soil, *J. Soil Sci.* 38: 453–459.

Baveye, P., M.B. McBride, D. Bouldin, T.D. Hinsley, M.S.A. Dahdoh and M.F. Abdel–Sabour, 1999, Mass balance and distribution of sludge-borne trace elements in a silt loam soil following long-term applications of sewage sludge, *Sci. Total Environ.* 227: 13–28.

Beckett, P.H.T., R.D. Davies and P. Brindley, 1979, The disposal of sewage sludge onto farmland: The scope of the problems of toxic elements, *Water Pollut. Control* 78: 419–445.

Berti, W.R. and L.W. Jacobs, 1996, Chemistry and phytotoxicity of soil trace elements from repeated sewage sludge applications, *J. Environ. Qual.* 25: 1025–1032.

Bidwell, A.M. and R.H. Dowdy, 1987, Cadmium and zinc availability to corn following termination of sewage sludge application, *J. Environ. Qual.* 16: 438–442.

Bingham, F.T., A.L. Page, R.J. Mahler and T.J. Ganje, 1975, Growth and cadmium accumulation of plants grown on a soil treated with cadmium-enriched sewage sludge, *J. Environ. Qual.* 4: 207–211.

Brown, S.L., R.L. Chaney, C.A. LLoyd, J.S. Angle, and J.A. Ryan, 1996, Relative uptake of cadmium by garden vegetables and fruits grown on long-term biosolids-amended soils, *Environ. Sci. Technol.* 30(12): 3508–3511.

Brown, S.L., R.L. Chaney, J.S. Angle and J.A. Ryan, 1998, The phytoavailability of cadmium to lettuce in long-term biosolids-amended soils, *J. Environ. Qual.* 27: 1071–1078.

CAST, 1976, Application of sewage sludge to cropland: Appraisal of potential hazards of the heavy metals to plants and animals, Council for Agricultural Science and Technology, Report 83, Ames, IA.

CAST, 1980, Effects of sewage sludge on the cadmium and zinc content of crops, Council for Agricultural Science and Technology, Report No. 83, Ames, IA.

Chaney, R.L., 1973, Crop and food chain effects of toxic elements in sludges and effluents, pp. 129–143, Conf. on Recycling Municipal Sludges and Effluents on Land, Champaign, IL, National Association of State Universities and Land-Grant Colleges.

Chaney, R.L., 1975, Metals in plants — absorption mechanisms, accumulation and tolerance, pp. 79–96, *Metals in Biosphere*, University of Guelph, Ontario.

Chaney, R.L., 1980, Health risks associated with toxic metals in municipal sludge, 59–83, in G. Bitton, B.L. Damron, G.T. Edds and J.M. Davidson (Eds.), *Sludge–Health Risks of Land Application*, Ann Arbor Science Publ., Ann Arbor, MI.

Chaney, R.L., 1983, Potential effects of waste constituents on the food chain, pp. 152–240, J.F. Parr, P.B. Marsh and J.M. Kla (Eds.), *Land Treatment of Hazardous Wastes*, Noyes Data Corp., Park Ridge, NJ.

Chaney, R.L. and J.A. Ryan, 1992, Regulating residual management practices, *Water Environ. Tech.* 4: 36–41.

Chaney, R.L., 1994, Trace metal movement: Soil–plant systems and bioavailability of biosolids-applied metals, C.E. Clapp, W.E. Larson and R.H. Dowdy (Eds.), *Sewage Sludge: Land Utilization and the Environment*, American Society of Agronomy, Madison, WI.

Chang, A.C., A.L. Page, K.W. Foster and T.E. Jones, 1982, A comparison of cadmium and zinc accumulation by four cultivars of barley grown in sludge-amended soils, *J. Environ. Qual.* 11(3): 409–412.

Chang, A., J.E. Warneke, A.L. Page and L.J. Lund, 1984, Accumulation of heavy metals in sewage sludge-treated soils, *J. Environ. Qual.* 13: 87–91.

Chang, A.C., A.L. Page and J.E. Warneke, 1987, Long–term sludge applications on cadmium and zinc accumulation in Swiss chard and radish, *J. Environ. Qual.* 16: 217–221.

Chang, A.C., H. Hyun and A.L. Page, 1997, Cadmium uptake for Swiss chard grown on composted sewage sludge treated field plots: Plateau or time bomb? *J. Environ. Qual.* 26: 11–19.

Chaudri, A.M., M.G.A. Celine, S.H. Badawy, M.L. Adams, S.P. McGrath and B.J. Chambers, 2001, Cadmium content of wheat grain from long-term field trial experiment with sewage sludge, *J. Environ. Qual.* 30(5): 1575–1580.

Corey, R.B., R. Fujii and L.L. Hendrickson, 1981, Bioavailability of heavy metals in sludge–soil systems, *Proc. 4th Annual Madison Conf. Appl. Res. Pract. Munic. Ind. Waste*, University of Wisconsin–Extension, Madison, WI.

Corey, R.B., L.D. King, C. Lue-Hing, D.S. Fanning, J.J. Street and J.M. Walker, 1987, Effects of sludge properties on accumulation of trace elements by crops, pp. 25–51, A.L. Page et al. (Eds.), *Land Application of Sludge*, Lewis, Chelsea, MI.

Dowdy, R.H. and V.V. Volk, 1983, Movement of heavy metals in soils, 229–240, in D.W. Nelson, K.K. Tanji and D.E. Elrick (Eds.), *Chemical Mobility and Reactivity in Soil Systems*, Soil Science Society of America, Madison, WI.

Dowdy, R.H., J.J. Ltterell, T.D. Hinesly, R.B. Grossman and D.L. Sullivan, 1991, Trace metal movement in an Aeric Ochraqualf following 14 years of annual sludge applications, *J. Environ. Qual.* 20: 119–123.

Dowdy, R.H., S.E. Clapp, D.R. Linden, W.E. Larson, T.R. Halbach and R.C. Polta, 1994, Twenty years of trace metal partitioning on the Rosemont sewage sludge watershed, 149–155, C.E. Clapp, W.E. Larson and R.H. Dowdy (Eds.), *Sewage Sludge: Land Utilization and the Environment*, American Society of Agronomy, Madison, WI.

Epstein, E., 1971, Effect of soil temperature on mineral element composition and morphology of the potato plant, *Agronomy J.* 63: 664–666.

Epstein, E. and R.L. Chaney, 1978, Land disposal of toxic substances and water–related problems, *J. Water Pollut. Control Fed.* 50(8): 2037–2042.

Giordano, P.M., D.A. Mays and A.D. Behel, Jr., 1979, Soil temperature effects on uptake of cadmium and zinc by vegetables grown on sludge–amended soil, *J. Environ. Qual.* 8: 233–236.

Hinsley, T.D., D.E. Alexander, K.E. Redborg and E.L. Ziegler, 1982, Differential accumulations of cadmium and zinc by corn hybrids grown on soil amended with sewage sludge, *Agron. J.* 74: 469–474.

Jenne, E.A., 1968, Controls on Mn, Fe, Co, Ni, Cu and Zn concentrations in soils and water. The significant role of hydrous Mn and Fe oxides, 337. *Trace Inorganics in Water*, Advan. Chem Ser., 73. American Chemical Society, Washington, D.C.

Jing, J. and T.J. Logan, 1992, Effects of sewage sludge cadmium concentration on chemical extractability and plant uptake, *J. Environ. Qual.* 21: 73–81.

Johnston, A.E., 1975, The Woburn market garden experiment: II. The effects of the treatments on soil pH, soil carbon, nitrogen, phosphorus and potassium, Rep. Rothamsted Experimental Station for 1974, Part 2: pp. 102–131, Rothamsted, U.K.

Johnston, A.E. and R.W.M. Wedderburn, 1975, The Woburn market garden experiment, 1942–1969: I. A history of the experiment, details of the treatments and yields of the crop. Rep. Rothamsted Experimental Station for 1974, Part 2: 79–101. Rothamsted, U.K.

Kim, S.J., A.C. Chang, A.L. Page, and J.E. Warneke, 1988, Relative concentration of cadmium and zinc in tissues of selected food plants grown on sludge-treated soils, *J. Environ. Qual.* 17 (4): 568–573.

Kochian, L.V., 1991, Mechanisms of micronutrient uptake and translocation in plants, pp. 229–296, J.J. Mortvedt, F.R. Cox, L.M. Shuman and R.M. Welch (Eds.), *Micronutrients in Agriculture*, Soil Science Society of America, Madison, WI.

Lake, D.L., P.W.W Kirk and J.N. Lester, 1984, Fractionation, characterization and speciation of heavy metals in sewage sludge and sludge-amended soils: A review, *J. Environ. Qual.* 13(2): 175–183.

Lindsay, W.L., 1979, *Chemical Equilibria in Soils*, John Wiley & Sons, New York.

Logan, T.J., 1997, Balancing benefits and risks in biosolids, *BioCycle* 38(1): 52–57.

Logan, T.J. and R.L. Chaney, 1983, Utilization of municipal wastewater and sludges on land-metals, in Workshop on Utilization of Municipal Wastewater and Sludge on Land, Denver, CO.

McBride, M.B, 1985, Sorption of copper (II) on aluminum hydroxide as affected by phosphate, *Soil Sci. Soc. Am. J.* 49: 843–846.

McBride, M.B, 1995, Toxic metal accumulation from agricultural use of sludge: Are USEPA regulations protective? *J. Environ. Qual.* 24: 5–18.

Mortvedt, J.J., F.R. Cox, L.M. Shuman and R.M. Welch, 1991, *Micronutrients in Agriculture*, Soil Science Society of America, Madison, WI.

O'Conner, G.A., 1988, Use and misuse of the DTPA soil test, *J. Environ. Qual.* 17: 715–718.

O'Connor, G.A., R.B. Brobst, R.L. Chaney, R.L. Kincaid, L.R. McDowell, G.M., Pierzynski, A. Rubin and G.G. Van Riper, 2001, A modified risk assessment to establish molybdenum standards for land application of biosolids, *J. Environ. Qual.* 30: 1490–1507.

O'Conner, G.A., T.C. Granato and R.H. Dowdy, 2001, Bioavailability of biosolids molybdenum to corn, *J. Environ, Qual.* 30: 140–146.

Page, A.L., T.J. Logan and J.A. Ryan, 1987, *Land Application of Sludge*, Lewis, Chelsea, MI.

Quirk, J.P. and A.M. Posner, 1975, Trace element adsorption by soil minerals, 95. *Trace Elements in Soil–Plant–Animal Systems*, Academic Press, Washington, D.C.

Sloan, J.J., R.H. Dowdy, M.S. Dolan and D.R. Linden, 1997, Long–term effects of biosolids application on heavy metal bioavailability in agricultural soils, *J. Environ. Qual.* 26: 966–974.

Sloan, J.J., R.H. Dowdy and M.S. Dolan, 1998, Recovery of biosolids–applied heavy metals sixteen years after application, *J. Environ. Qual.* 27: 1312–1317.

Sposito, G., L.J. Lund and A.C. Chang, 1982, Trace metal chemistry in arid–zone field soils amended with sewage sludge: I. Fractionation of Ni, Cu, Zn, Cd and Pb in solid phases, *Soil Sci. Am. J.* 46: 260–264.

Sposito, G., C.S. LeVesque, J.P. LeClaire and A.C. Chang, 1983, Trace metal chemistry in arid–zone field soils amended with sewage sludges: III. Effect of time on the extraction of trace metals, *Soil Sci. Soc. Am. J.* 47: 898–902.

USEPA, 1980, Sewage sludge: Factors affecting the uptake of cadmium by food–chain crops grown on sludge–amended soils, W–124 SEA–CR Technical Res. Committee, U.S. Environmental Protection Agency Report No. SW–882, Washington, D.C.

Ward, G.M., 1994, Molybdenum requirements, toxicity and nutritional limits for man and animals, pp. 452–476, E.R. Braithwaite and J. Haber (Eds.), *Molybdenum: An Outline of its Chemistry and Use*, Studies in Inorganic Chemistry 19, Elsevier, Amsterdam.

Williams, D.E., J. Vlamis, A.H. Pukite and J.E. Corey, 1987, Metal movement in sludge–amended soils: A nine–year study, *Soil Sci.* 143: 124–131.

CHAPTER **6**

Organic Chemicals

INTRODUCTION

Both natural and xenobiotic (manmade) organic compounds abound in the universe. Many of these compounds are toxic to humans and animals. Gribble (1994) points out that numerous chlorinated compounds are naturally produced. These include organohalogens, numerous halogenated alcohols, ketones, carboxylic acids and amides, aldehydes, epoxides and alkenes. Many of these are produced by fungi and marine algae, as well as during volcanic action, forest fires and brush and vegetation burning. He indicates that nearly 100 different chlorinated, brominated and iodinated compounds have been found in an edible seaweed favored by Hawaiians.

Our environment has been greatly contaminated by toxic organic chemicals, primarily as a result of industrial discharges and uses of pesticides. Industrial and manufacturing enterprises produce a myriad of organic chemicals. It has been estimated that more than 5 million distinct organic compounds are registered.

Following World War II, pesticide and herbicide usage in agriculture increased dramatically and many of the compounds used were very persistent in the environment. Kuhn and Suflita (1989) indicate that in recent years, 17 pesticides have been found in groundwater in 23 states. In the 1960s, there was increased awareness of the potential harmful effects of many of these organic compounds on humans, fish and wildlife. Through soil, water, or air, many of those compounds enter the sewer system and end up in biosolids.

Biosolids can contain toxic organic chemicals, principally as discharges from industrial sources, but also from atmospheric deposition (Webber and Lesage, 1989; USEPA, 1990; Jones and Sewart, 1997). Today many manufacturers and producers of organic compounds pretreat their wastewater prior to discharging into the sewer system. When organic compounds enter the wastewater treatment system, they can undergo reductions or transformations prior to being deposits in biosolids that will be applied to land. For example, chlorinated aliphatic compounds can undergo reductive dechlorination, hydrolysis, dehydrochlorination and dihaloelimination

87

(Ballapragada et al., 1998). The authors found that specific bacteria accomplished dechlorination of several chlorinated organic compounds.

Jones and Sewart (1997) speculated that the process of potential importance during wastewater treatment could involve (1) deposition of dioxins in the biosolids, (2) microbial degradation during digestion, (3) volatilization of the lower-chlorinated homologue groups and (4) possible formation of PCDD/F during the wastewater treatment by microbial mediated reactions. This chapter does not discuss the fate of organic compounds during wastewater treatment.

The data on organic compounds in sewage sludge and biosolids were presented in Chapter 2, Characteristics of Sewage Sludge and Biosolids.

The organic compounds of greatest concern are:

- Toxic chlorinated compounds
- Alkylphenol ethoxylates
- Volatile organic compounds (VOCs)
- Dioxin and dioxin-like compounds
- Phthalates
- Polycyclic aromatic hydrocarbons (PAHs)
- Pesticides

The chlorinated compounds of major concern are polychlorinated biphenyls (PCBs). It is estimated that since 1935, more than 60,000 metric tonnes have been produced in the U.S. PCBs are very persistent and bioaccumulate. PCBs in soils can be taken by plants (Strek and Weber, 1980; O'Conner et al., 1990).

Alkylphenol ethoxylates are nonionic surfactants. The main alkylophenols used are nonylphenol ethoxylates. Alkylphenol ethoxylates compounds can be microbially metabolized (Ahel et al., 1994). Nonylphenol ethoxylates are susceptible to photochemical degradation.

Some of the sources of several priority pollutants are shown below:

- PCBs — electrical capacitors and transformers, paints, plastics, insecticides
- Halogenated aliphatics — fire extinguishers, refrigerants, propellants, pesticides, solvents
- Phthalate esters — polyvinyl chloride and thermoplastics
- Ethers — solvents for polymer plastics
- Phenols — synthetic polymers, dyestuffs, pigments, pesticides, herbicides
- Polycyclic aromatic hydrocarbons — dyestuffs, pesticides, herbicides, motor fuels and oils
- Dioxins — herbicides, pulp bleaching, emissions from waste incinerators, textiles
- Nonylphenol ethoxylates (NPEs) — pulp and paper, plastics, household cleaning agents, pesticides and paint

Consideration must be given both to the risk from a pollutant and to the potential for exposure. Exposure is a function of the presence of the compound in the media in question — in this case biosolids — but also to its presence in other sources. In fact, many times, exposure from food, air and water may greatly outweigh any potential exposure from biosolids applied to land. Chaney et al. (1990) estimated the risk from PCBs in sewage sludge applied to soil. They concluded that since sewage sludges contained very low levels of PCBs, the estimated risk level to the Most Exposed Individuals (MEIs) was <10^{-4}; low sludge PCBs and low probability of MEIs are at <10^{-7} lifetime risk.

Furthermore, since untreated sewage sludge is not permitted to be applied to agricultural land, the additional transformation to biosolids can significantly reduce the level of organics. Chlorinated organic compounds can be biodegraded during anaerobic digestion (Ballapragada et al., 1998). During composting many organics are biodegraded; heat drying will destroy or volatilize many organics; and alkaline treatment will also transform or destroy some. There have been numerous studies showing that composting biodegrades many toxic organic compounds (Rose and Mercer, 1968; Deever and White, 1978; Snell Environmental Group Inc., 1982; Epstein, 1997; Laine and Jorgensen, 1997; Cole, 1998; Potter et al., 1999).

When biosolids containing toxic organic compounds are applied to land, the compounds can undergo numerous transformations and reactions. These can affect their potential impact to humans, animals and the environment. The main impact on humans and animals is through the food chain and water intake. Jones and Sewart (1997) list four major pathways for organic contaminants transfer into humans.

- Biosolids to soil to root crops
- Biosolids to soil to above ground crops
- Biosolids to soil to livestock ingestion to milk and animal tissues
- Biosolids to soil to groundwater to drinking water

They indicate that there are possibly other unusual and relatively minor pathways, such as direct ingestion of soil containing biosolids.

The main objective of this chapter is to provide information on the fate and potential impact of organic compounds in biosolids when applied to soil. Several excellent sources of more detailed information include reviews by Jones and Sewart (1997); Alexander (1995); Chaney et al. (1996); Sawhney and Brown (1989); and the Final Report by the Water Environment Association of Ontario (2001) entitled Fate and Significance of Selected Contaminants in Sewage Biosolids Applied to Agricultural Land Through Literature Review and Consultation with Stakeholder Groups.

FATE OF TOXIC ORGANIC COMPOUNDS WHEN BIOSOLIDS ARE LAND APPLIED

Soils throughout the world are contaminated with toxic organic compounds. Soil contamination can result from atmospheric deposition, combustion of wastes, waste disposal, spillage of industrial materials, usage of pesticides, discharges of household chemicals and numerous other ways. For example, atmospheric deposition is the primary means of soil contamination of PCDD/Fs, according to Jones and Sewart (1997). They reported that soils in urban and industrial areas had higher concentrations than rural soils.

Table 6.1   Level of Selected Toxic Organic Compounds in Rural North American Soils

Compound	Number of Samples	Geometric Mean g Mg^{-1}	Standard Deviation
PCDD/Fs*	70	0.46×10^{-6}	5.1
PCBs	1,483	0.007	2.7
PAHs	>24	0.06	4.3

* The authors quantified PCDD/Fs in terms of I-TEQ values that were estimated from published homologue and congener data.

Source: Cook and Beyea, 1998, Toxicol. Environ. Chem. 67: 27–69. With permission.

Table 6.2   Estimated Atmospheric Deposition of Selected Toxic Organic Compounds

Compound	Number of Locations	Mean g ha^{-1} yr^{-1}
PCDD/Fs*	>4	16×10^{-6}
PCBs	9	0.1
PAHs	6	3.5

* The authors quantified PCDD/Fs in terms of I-TEQ values that were estimated from published homologue and congener data.

Source: Cook and Beyea, 1998, Toxicol. Environ. Chem. 67: 27–69. With permission.

No systematic survey of dioxin soil levels has been conducted in the United States (USEPA, 1999). Based on the examination of data from numerous projects, USEPA reported that, for rural soils, the values ranged from one to six ppt I-TEQ (International Toxic Equivalents). In Table 6.1, Cook and Beyea (1998) showed levels of several organic compounds in rural North American soils.

Cook and Beyea (1998) also published data on atmospheric deposition of selected organic compounds (see Table 6.2) and time constants for the disappearance of these compounds (see Table 6.3). Apparently, these compounds are very persistent in soils.

When biosolids are land applied, toxic organic compounds can undergo numerous reactions and transformations. These can affect their movement through the soil to water resources, uptake by plants, volatilization to the atmosphere, accumulation in soil biota and other fates. The major reactions, transformations and fate in the soil are:

- Volatilization and wind erosion
- Photodecomposition or photochemical degradation
- Foliar interception and adsorption
- Plant uptake and crop removal
- Runoff and erosion to surface waters
- Soil adsorption and desorption
- Leaching to groundwater
- Biological degradation
- Chemical decomposition
- Uptake by soil biota

Table 6.3   Time Constants for the Disappearance of Selected Toxic Organic Compounds
from Soils Treated with Biosolids

Compound	Number of Years of the Study	% Loss	Estimated Number of Years for Disappearance	Reference
PCDD/Fs	22	26–50	>100	McLachlan et al. (1996)
PCBs	0.66	8–33	2.5–5.9	Fairbanks et al. (1987)
	30	91	14–19	Alcock et al. (1996)
PAHs	30	39-45	30	Wild et al. (1990)
	21	ca. 90	9	Wild et al. (1991)

In a review entitled "How Toxic Are Chemicals in Soil?" Alexander (1995) indicates that the hazard and risk from toxic chemicals diminish as the compounds persist in soil. Many of the reactions and transformations are responsible for the diminishing of organic chemicals.

Most of the scientific literature dealing with persistence and fate of organic chemicals pertained to pesticides. However, in recent years, considerable focus has been given to the chlorinated compounds. These compounds are often very persistent and remain in soils for long periods of time. The half-life of PCDD/Fs in surface soils may be on the order of 10 years or more (Alcock et al., 1996). USEPA indicates that there is no systematic survey of dioxin soil levels in the United States (USEPA, 1999). Rural soils values generally range from 1 to 6 ppt I-TEQ (International Toxic Equivalents) and urban soil values range from seven to 20 I-TEQ.

Volatilization

Volatilization of organic compounds from soil can be very significant. An organic compound in biosolids, spread on the surface or incorporated into the soil, will partition between the gas and liquid phases to exert a vapor pressure. This vapor may be rapidly lost. Many aromatics, such as toluene, benzene, cyclohexane and others, are volatile. Increasing the organic matter level of soil tends to decrease the volatilization of the hydrophobic nonpolar aromatics.

Fairbanks et al. (1987) reported that in an unamended soil, volatilization of PCBs ranged from 5% to 31%. Volatilization of organics was the major means of loss of ^{14}C in unamended soil. The addition of biosolids decreased the rate of volatilization and environmental transport. They indicated that sewage sludge can decrease plant uptake of PCB since foliar contamination from vapor sorption is the primary source of PCB contamination of some plants. Furthermore, the addition of sewage sludge increased the complete degradation or detoxification of PCBs.

Jin and O'Conner (1990) found in a laboratory study that more than 80% of toluene was volatilized from either unamended or sludge-amended soils. They indicated that volatilization is a major force in toluene movement. Therefore, to reduce possible groundwater and air pollution by toluene due to sludge application, the sludge should be incorporated into the surface layer or deeper.

Wilson and Jones (1999) reported that volatilization was the predominant loss process for volatile organic compounds (VOCs) from land application of biosolids.

The rate of loss depended on biosolids' application rate, method of application, soil properties and compound characteristics. When biosolids are applied to land, the loss of VOCs, PCBs and chlorophenols (CPs) by volatilization can be important (Wilson et al., 1997). Dioxins and furans may be volatilized and adsorbed by foliage. Carpenter (2000) indicates that the half-lives for PCDD/Fs range from 10 to 17 years, but when residues are on the surface the half-life is considerably shorter. This is probably due to both volatilization and photo-oxidation.

Photodecomposition

There is virtually no data on photodecomposition of organic compounds from surface-applied biosolids. Some organic compounds in biosolids applied to the soil surface can undergo photodecomposition. Phenolics and polynuclear compounds can undergo such reactions when exposed to solar radiation. Usually photodecomposition is measured as part of the total decomposition by biological and abiotic mechanisms.

Degradation

Many bacteria, fungi and other organisms have been found to degrade organic compounds under aerobic conditions. For example, Saber and Crawford (1985) isolated strains of *Flavobacterium* that degraded pentachlorophenol (PCP). The white rot fungi have been found to degrade a wide host of organic compounds. Barr and Aust (1994) listed the environmental pollutants degraded by the white rot fungus *Phanerochaete chrysosporium*. These include:

- Chlorinated aromatic compounds
  - Pentachlorophenol (PCP)
  - 2,4,5-Trichlorophenoxyacetic acid
  - Polychlorinated biphenyls (PCB)
  - Dioxin

- Polycyclic aromatic compounds
  - Benzo(a)pyrene
  - Pyrene
  - Anthracene
  - Chrysene

- Pesticides
  - 1,1,1-Trichloro-2,2-bis(4-chlorophenyl)ethane (DDT)
  - Lindane
  - Chlordane
  - Toxaphene

Many organic compounds will degrade in the soil. This process is extremely important as a means of removing several toxic organic compounds. There is extensive literature on the degradation of pesticides in soil. Organic matter can either accelerate or possibly inhibit biodegradation. Guthrie and Pfaender (1998) determined that biodegradation was the main means of removing pyrene. Accelerated biodegradation could occur through the enhancement of the microbial population and its activity. Organic matter can also tie up compounds, thus reducing or delaying their assimilation by the microbial population. Adding compost to soils accelerates the degradation of toxic organic compounds. Stegmann et al. (1991), Atlas (1991), and Hupe et al. (1996) showed that compost added to soils was effective in hydrocarbon degradation.

Bellin et al. (1990) studied the degradation of pentachlorophenol (PCP) in biosolids-amended soils. Pentachlorophenol is highly toxic. It has been used primarily as a fungicide and insecticide in the preservation of wood. Extensive use of this compound has resulted in contamination of soils, water, air and biosolids (Buhler et al., 1973). DNP or 2,4-dinitrophenol is a compound toxic to animals and plants that occurs as a waste contaminant from several industrial sources (O'Conner et al., 1990). The authors found that it was rapidly degraded in soils and only slightly affected by biosolids. Degradation of PCP appeared to be more favorable in high-pH soils. Their data for the Norfolk soil suggested a first-order degradation with a half-life of about 38 days. Wilson et al. (1997) indicated that biodegradation was a very important process in the loss of toluene and o-, m- and p-xylene, PCBs and PCPs in soil. This was not true of PCDDs/Fs.

Nonylphenol ethoxylates (NPEs) are found in soils where biosolids have been applied. Marcomini et al. (1989) evaluated the fate of various NPEs in sludge-amended soil. They determined that 80% of NPEs in the sludge-amended soil degraded in the first month. There remained residual levels after 320 days. Possibly, some of the NPEs may have leached out of the analytical zone. NPE does not appear to significantly move through soils to groundwater (Langenkamp and Part, 2001).

Fairbanks and O'Connor (1982) showed that 84% to 89% of di (2-ethylhexyl) phthalate (DEHP) was degraded in 146 days. Overcash (1983) indicated that the decomposition half-life for di-n-butylphalate ester was approximately 80 to 180 days and for nonionic surfactants, 300 to 600 days.

Kuhnt (1993) reported that surfactants used in domestic detergents such as LAS and non-ionic LAE are rapidly and extensively degraded in biosolids-amended soil and even in soils with no previous exposure to those compounds. Managas et al. (1998) also reported complete loss of LAS from biosolids-amended soil within 98 to 336 days. Under aerobic conditions, LAS has been reported to degrade rapidly (Litz et al., 1987; Madsen et al., 1997). Jensen (1999) concluded that the combination of relatively rapid aerobic degradation and reduced bioavailability when biosolids are applied likely prevents LAS from posing a threat to terrestrial ecosystems on a long-term basis.

PLANT UPTAKE OF ORGANIC COMPOUNDS

The potential plant uptake of toxic organic compounds from biosolids-amended soils depends on (a) the presence of the compound and its concentration in the biosolids; (b) the chemical and physical properties of the compound; (c) reactions in the soil that affect its availability and (d) rate of uptake by plants. Simonich and Hites (1995) determined that environmental factors and plant species are also important factors. As was indicated in Chapter 2, biosolids generally contain very low levels of toxic organic compounds. The principal concern has been with low levels of highly persistent and toxic compounds such as PCBs, PAHs, dioxin, PCPs, phthalates and similar compounds.

Bioconcentration factors (BCFs) are used to quantify plant contamination. Organic chemicals can enter the plant from a contaminated soil and be translocated in the plant through the xylem. The compound needs to be soluble, since the xylem transports water from the roots to the leaves by transpiration. Organic compounds can also enter through the leaves from the atmosphere and be translocated by the phloem. These pathways are a function of (a) the chemical and physical nature of the compound, such as lipophilicity and water solubility; (b) environmental factors such as ambient temperature; (c) edaphic factors such as organic content of the soil; and (d) plant species (Simonich and Hites, 1995).

Early data demonstrated that several organochlorine pesticides are adsorbed by root crops (Lichtenstein, 1959; Lichtenstein and Schulz, 1965; Harris and Sans, 1967; Beall and Nash, 1971). Iwata and Gunther (1976) reported that carrot roots absorbed PCBs. However, 97% of the PCBs were found in the peel, with very little translocated in the plant tissue.

Studies on PCB uptake from biosolids-amended soils are limited. Davis et al. (1981) showed that with Milorganite, a heat-dried biosolids product containing low concentrations of PCBs (20 to 40 mg/kg), there was no detectable PCB in grass. O'Connor et al. (1990) studied the uptake of PCBs by fescue, carrot and lettuce from a highly contaminated sludge. The sludge contained 52 mg/kg of PCBs. Only carrots were contaminated, though the PCB contamination was restricted to the peel. The concentration of PCBs in this study was considerably higher than concentrations of one to five mg/kg typically found in biosolids.

Ingestion by animals of plants and accumulation in animal byproducts, represent another mechanism for a toxic organic compound to enter the food chain. Extensive studies were done on PCB intake to tissue and milk in dairy cattle (Fries et al., 1973; Willet, 1975). They showed that Aroclor 1254 was concentrated in milk fat.

Fries (1982) cites three means of plant contamination: direct, indirect and soil ingestion by animals. Direct plant contamination occurs when biosolids are applied to a crop and adheres to plant surfaces. In the case of liquid sludge application to forage crops, the PCB intake by animals will depend on whether they are allowed to graze shortly after application. Currently, the 503 regulations require a waiting period of 30 days prior to grazing. However, if the forage is mowed and fed to animals, some of the biosolids would adhere to the crop. Indirect intake by animals depends on the root adsorption and translocation to the foliage. As indicated earlier, scientific evidence shows that PCBs are not taken up by most plant roots and translocated to the aboveground portions of plants. Grazing animals will ingest soil. Fries (1982) found that grazing dairy cattle's ingested soil represents as much as 14% of their dry-matter diet. Thus, biosolids containing toxic organics, when left on the soil surface, can be ingested along with the soil. Incorporating biosolids into the soil is the best way to reduce animal exposure.

PAHs are ubiquitous and are found in soils and foliage due to atmospheric deposition (Wild and Jones, 1992). Plant uptake through the roots is limited since organic matter adsorbs them. The organic matter in biosolids increases the adsorption potential, thus making them less available to plants (Ryan et al., 1988). Furthermore, since PAHs are lipophilic/hydrophobic compounds, they tend to be more greatly adsorbed by soil organic matter.

Wild and Jones (1992) suggest that root uptake may be enhanced by the presence of surfactants in biosolids and that adsorption onto root surfaces can be an important process in root uptake. They analyzed carrot foliage, root peels and root cores for 15 PAHs. Neither foliage nor root cores were affected by anaerobically digested biosolids. Carrot root peel PAH concentrations increased to a plateau with increasing soil PAH levels.

Pentachlorophenol (PCP), the compound used as a wood preservative, has contaminated water, air, food and sediment over the years (Bevenue and Beckman, 1967; Buhler et al., 1973). Bellin and O'Connor (1990) evaluated the uptake of PCP by tall fescue, lettuce, carrot and chili pepper from biosolids-amended soil. They found minimal plant uptake of intact PCP in fescue and lettuce and none in carrot or chile plants. Plant dry wt/initial soil concentration (BCF values) for the lettuce and fescue were <0.01. They found that degradation in the soil was rapid and that minimal contamination could occur in the field.

Chlorobenzenes (CBs) can be found in biosolids. They are lipophilic and volatile and therefore can be taken up by both the roots and foliage (USEPA, 1985). Overcash et al. (1986) indicated that monochlorobenzene and 1,4-dichlorobenzene reached the highest BCFs into crops grown on sludge-amended soil. Wang and Jones (1994) studied the uptake of CBs by carrots grown in soil treated with different rates of sewage sludge. Both carrot foliage and roots took up CBs from all the soil treatments. There was no evidence that the CBs were translocated from the roots to the tops. There was some penetration by dichlorobenzene from the peel to the core.

Phthalates are found in sludge and biosolids. The most common compound is di-(2-ethylhexyl) phthalate (DEHP). Studies prior to 1990 investigated the uptake of DEHP from soil (Overcash et al., 1986). DEHP is strongly adsorbed to the soil organic matter; therefore the presence of biosolids may decrease DEHP availability and reduce uptake by plants (Aranda et al., 1989). The authors studied the uptake of DEHP from sludge-amended soil by three food chain crops — lettuce, carrot and chili pepper — as well as tall fescue. Intact DEHP was not detected in any of the plants. The low bioconcentration factors suggested very little uptake of DEHP.

Since dioxins are considered highly toxic and are ubiquitous as a result of atmospheric deposition, recent attention has been focused on potential uptake by plants (Hülster et al., 1994; Welsch-Pausch et al., 1995). Jones and Sewart (1997) reviewed the uptake of PCDD/F by plants and the various pathways. They state that PCDD/F concentration in the aboveground portion of plants is controlled by foliar uptake and is virtually unaffected by changes in soil concentrations. Jones and Sewart (1997) and Wild et al. (1994) conclude that the influence of biosolids' application on PCDD/F concentration on aboveground plant tissues can be ignored in the pathway analysis.

Table 6.4   Examples of Organic Compounds and the Organisms That Degrade Them Under Aerobic Conditions

Organic Compound	Organism	Environment
2-Chlorobenzoic acid	*Pseudomonas*	Aerobic, Soil
4-Chlorobenzoic acid	*Arthobacter*	Aerobic
3,5-Dichlorobenzoic acid	*Pseudomonas*	Aerobic
3-Chlorobenzene	*Pseudomonas*	Aerobic
1,4-Dichlorobenzene	*Pseudomonas*	Aerobic
1,4-Dichlorobenzene chlorobenzene	*Alcaligenes*	Aerobic
3-Chlorophenol	*Nocardia*	Aerobic
4-Chlorophenol	*Mycobacterium*	Aerobic
	Alcaligenes	
	Flavobacter	
Pentachlorophenol (PCP)	*Arthrobacter*	Soil
	Flavobacter	Aerobic
	Pseudomonas	Aerobic
	Coryneform	Aerobic
2,4-Dichlorophenoxyacetic acid (2,4-D)	*Pseudomonas*	Soil
	Azobacter	Aerobic
2,4,5-Trichlorophenoxacetic acid (2,4,5-T)	*Pseudomonas*	Soil
1,1,1-Trichloro-2,2-bis(*p*-chlorophenol)ethane (DDT)	*Escherica*	Aerobic
	Pseudomonas	
	Aerobacter	
	Clostridium	
	Proteus	
	Fusarium	
	Mucor	
	Cylindrotheca	
	Nocardia	
	Streptomycetes	
	Phanerochaete	
Polychlorobiphenyl (PCB)	*Alcaligenes*	Aerobic, soil
	Acintobacter	Aerobic
	Pseudomonas	Soil

Source: Boyle, 1989, *J. Environ. Qual.* 18(4): 395–402. With permission.

Hülster et al. (1994) reported that the fruits of zucchini had higher concentrations of PCDD/PCDF than other fruits and vegetables. They indicated that uptake by zucchini and pumpkin was principally from the roots with subsequent translocation to the shoots and fruit. However, as Jones and Sewart (1997) indicate, these crops make up a very small amount of our diet and therefore are not very significant as a human exposure pathway.

Cucumber plants were mainly contaminated by atmospheric deposition. In a study with Welsh ray grass, an important food chain plant, Welsch-Pausch et al. (1995) did not find that soil-related uptake was important in plant uptake of PCDD/PCDFs. Dry gaseous deposition was the principal pathway of contamination.

Table 6.4 presents some examples of organic compounds and organisms that degrade them under aerobic conditions. Thus, over time, many of the persistent organic compounds will biodegrade in the soil.

O'Connor (1998) found three factors that reduce the potential for plant contamination: low concentration of toxic organic compounds in biosolids-amended soil;

strong adsorption of toxic organics to soil or biosolids; and loss of toxic organics from soil through degradation.

Thus, O'Connor (1998) and Chaney et al. (1996) concluded that the potential uptake of biosolids–borne and compost-borne toxic organics is small. Furthermore, toxic organic compounds are very large molecules; therefore, with the exception of some root crops, they are not taken up by plants and translocated to the edible portion. Also, toxic organics may be metabolized within the plant. Jacobs et al. (1987) did not find any phytotoxic effects of organic pollutants when biosolids were land applied.

CONCLUSION

Both natural and xenobiotic (manmade) organic compounds abound in the universe. Many of these compounds are toxic to humans and animals. When organic compounds in biosolids are land applied, they may undergo numerous reactions and transformations in the soil. Surface application of biosolids can result in volatilization and photochemical degradation. In the soil the compounds may be chemically decomposed, adsorbed on soil particles, biodegraded, or move through the soil. Plant uptake of organic chemicals is minimal.

Foliar interception and adsorption can also occur primarily from atmospheric deposition. Soil biota may take up organic chemicals. Soil erosion and runoff can contaminate water courses. The relative risk from organic chemicals in biosolids has been shown to be minimal as a result of their relative low levels and their many transformations, especially biological degradation.

The Water Environment Association of Ontario 2001 in its final report entitled, "Fate and Significance of Selected Contaminants in Sewage Biosolids Applied to Agricultural Land Through Literature Review and Consultation with Stakeholder Groups" provided a synopsis of the properties, occurrence, fate and transfer of the principal organic contaminant groups found in sludge and biosolids. The information was based on a paper by Smith (1996).

REFERENCES

Ahel, M., D. Hrsak and W. Giger, 1994, Aerobic transformation of short-chain alkylphenol polyethoxylates by mixed bacterial cultures, *Arch. Environ. Contam. Toxicol.* 26: 540–548.

Alcock, R.E., J. Bacon, R.D. Bardgett, A.J. Beck, P.M. Hagarth, R.G.M. Lee, C.A. Parker and K.C. Lones, 1996, Persistence and fate of PCBs in sewage sludge agricultural soil, *Environ. Pollut.* 93: 83–92.

Alexander, M, 1995, How toxic are toxic chemicals in soil? *Environ. Sci. Technol.* 29(11): 2713–2717.

Aranda, J.M., G.A. O'Conner and G.A. Eiceman, 1989, Effects of sewage sludge on di-(2-ethylhexyl)phthalate uptake by plants, *J. Environ. Qual.* 18: 45–50.

Atlas, R.M., 1991, Bioremediation of fossil fuel contaminated soils, p. 15, in R.E. Hinchee and R.F. Olfenbuttel (Eds.), *In Situ Bioremediation*, Battelle, Columbus, OH.

Ballapragada, B., H.D. Stensel, J.F. Ferguson, V.S. Magar and J.A. Puhakka, 1998, Toxic chlorinated compounds: Fate and biodegradation in anaerobic digestion, Water Environment Research Foundation, Project 91-TFT-3, Alexandria, VA.

Barr, D.P. and S.D. Aust, 1994, Mechanisms white rot fungi use to degrade pollutants, *Environ. Sci. Technol.* 28(2): 78A–87A.

Beall, M.L. and R.G. Nash, 1971, Organochlorine insecticide residues in soybean plant tops: Root vs. vapor sorption, *Agron. J.* 63: 460–464.

Bellin, C.A., G.A. O'Conner and Y. Jin, 1990, Sorption and degradation of pentachlorophenol in sludge amended soil, *J. Environ. Qual.* 19: 603–608.

Bevenue, A. and H. Beckman, 1967, Pentachlorophenol: A discussion of its properties and its occurrence as a residue in human and animal tissues, *Res. Rev.* 19: 83–144.

Boyle, M., 1989, The environmental microbiology of chlorinated aromatic decomposition, *J. Environ. Qual.* 18(4): 395–402.

Buhler, D.R., M.E. Rassmusson and H.S. Nakaue, 1973, Occurrence of hexachlorobenzene and pentachlorophenol in sewage sludge and water, *Environ. Sci. Technol.* 7: 929–934.

Carpenter, A., 2000, Dioxin in organic residues, 14th Annual Residuals and Biosolids Management Conference, Boston, Water Environment Research Foundation, Alexandria, VA.

Chaney, R.L., J.A. Ryan and G.A. O'Conner, 1990, Risk assessment for organic micropollutants, *Proc. EEC Symp. Treatment and Use of Sewage Sludge and Liquid Agricultural Wastes*, Athens, Greece.

Chaney, R.L., J.A. Ryan and G.A. O'Connor, 1996, Organic contaminants in municipal biosolids: Risk assessment, quantitative pathway analysis and current research priorities, *Sci. Total Environ.* 185(1–3): 187–216.

Cole, M.A., 1998, Bioremediation of soils contaminated with toxic organic compounds, pp. 175–194, S. Brown, J.S. Angle and L. Jacobs (Eds.), *Beneficial Co-utilization of Agricultural, Municipal and Industrial By–products*, Kluwer Academic, Dordrecht, Netherlands.

Cook, J. and J. Beyea, 1998, Potential toxic and carcinogenic chemical contaminants in source–separated municipal solid waste composts: Review of available data and recommendations, *Toxicol. Environ. Chem.* 67: 27–69.

Davis, R.D., J.L. Pyle, J.H. Skillings and N.D. Danielson, 1981, Uptake of polychlorobiphenyls present in trace amounts in dried municipal sewage sludge through an old field ecosystem, *Bull. Environ. Contam. Toxicol.* 27: 689–694.

Deever, W.R. and R.C. White, 1978, Composting petroleum refinery sludges, Texaco, Inc., Port Arthur, TX.

Epstein, E., 1997, *The Science of Composting*, Technomic, Lancaster, PA.

Fairbanks, B.C. and G.A. O'Conner, 1982, Fate of toxic organics in sludge–amended soils. *Proc. National Conference on Composting of Municipal and Industrial Sludges*, Washington, D.C., Hazardous Materials Control Research Institute.

Fairbanks, B.C., G.A. O'Conner and S.E. Smith, 1987, Mineralization and volatilization of polychlorinated biphenyls in sludge-amended soil, *J. Environ. Qual.* 16(1): 18–25.

Fries, G.F., 1982, Potential polychlorinated biphenyl residues in animal products from application of contaminated sewage sludges to land, *J. Environ. Qual.* 11 (1): 14–20.

Fries, G.F., G.S. Marrow and C.H. Gordon, 1973, Long–term studies of residue retention and excretions by cows fed polychlorinated biphenyl (Aroclor 1254), *J. Agric. Food Chem.* 21: 117–121.

Gribble, G.W., 1994, The natural production of chlorinated compounds, *Environ. Sci. Technol.* 28(7): 310A–319A.

Guthrie, E.A. and F.K. Pfaender, 1998, Reduced pyrene bioavailability in microbially active soils, *Environ. Sci. Technol.* 32: 501–508.

Harris, C.R. and W.W. Sans, 1967, Absorption of organochlorine insecticide residues from agricultural soils by root crops, *J. Agric. Food Chem.* 15 (861).

Hülster, A., J.F. Muller and H. Marschner, 1994, Soil–plant transfer of polychlorinated dibenzo-p-dioxins and dibenzofurans to vegetables of the cucumber family (Cucurbiataceae), *Environ. Sci. Technol.* 28: 1110–1115.

Hupe, K., J.C. Luth, J. Heerenklage and R. Stegmann, 1996, Enhancement of the biological degradation of contaminated soils by compost addition, pp. 913–923, M. de Bartoldi, P. Sequi, B. Lemmes and T. Papi (Eds.), *The Science of Composting*, Blackie Academic and Professional, London.

Iwata, Y. and F.A. Gunther, 1976, Translocation of the polychlorinated biphenyl aroclor 1254 from soil into carrots under field conditions, *Arch. Environ. Contam.* 4: 44–59.

Jacobs, L.W., G.A. O'Conner, M.R. Overcash, M.J. Zebeck and P. Rygwiecz, 1987, Effects of trace organics in sewage sludges on soil–plant systems and assessing their risks to humans, pp. 101–143, A.L. Page et al., (Eds.), *Land Application of Sludge — Health Effects*, Lewis Publishers, Chelsea, MI.

Jensen, J., 1999, Fate and effects of linear alkylbenzene sulphonates (LAS) in the terrestrial environment, *Sci. Total Environ.*, 226: 93–111.

Jin, Y. and G.A. O'Conner, 1990, Behavior of toluene added to sludge-amended soils, *J. Environ. Qual.* 19: 573–579.

Jones, K.C. and A.P. Sewart, 1997, Dioxins and furans in sewage sludges: A review of their occurrence and sources in sludge and of their environmental fate, behavior and significance in sludge-amended agricultural systems, *Crit. Rev. Environ. Sci. Technol.* 27(1): 1–85.

Laine, M.M. and K.S. Jorgensen, 1997, Effective and safe composting of chlorophenol–contaminated soil in pilot scale, *Environ. Sci. Technol.* 31(2): 371–378.

Langenkamp, H. and P. Part, 2001, Organic constituents in sewage sludge for agricultural use, European Commission Joint Research Centre, Institute for Environmental and Sustainability, Soil and Waste Unit.

Lichtenstein, E.P., 1959, Adsorption of some chlorinated hydrocarbon insecticides from soils into various crops, *J. Agric. Food Chem.* 7: 430.

Lichtenstein, E.P. and K.R. Schulz, 1965, Residues of aldrin and heptachlor in soils and their translocation into various crops, *J. Agric. Food Chem.* 13: 57.

Litz, N., H.W. Doring, M. Thiele and H.P. Blume, 1987, The behavior of LAS in different soils: A comparison between field and laboratory studies. *Ecotoxicol. Environ. Safety* 14: 103–116.

Kuhn, E.P. and J.M. Suflita, 1989, Dehalogenation of pesticides by anaerobic microorganisms in soil and groundwater — a review, B.L. Sawhney and K. Brown (Eds.), *Reactions and Movement of Organic Chemicals in Soils*, SSSA Special Pub. No. 22, Soil Science Society of America, Madison, WI.

Kuhnt, G., 1993, Behavior and fate of surfactants in soil, *Environ. Toxicol. Chem.* 12: 1813–1820.

Madsen, T., P. Kristensen, L. Samso-Petersen, J. Torslov and J.O. Rasmussen, 1997, Application of sludge on farmland — quality objectives, level of contamination and environmental risk assessment, Specialty Conference on Management and Fate of Toxic Organics in Sludge Applied to Land, Copenhagen.

Mangas, E., M.T. Vaquero, L. Comellas and F. Broto-Puig, 1998, Analysis and fate of aliphatic hydrocarbons, linear alkylbenzenes, polychlorinated biphenyls and polycyclic aromatic hydrocarbons in sewage sludge-amended soils, *Chemosphere* 1: 61–73.

Marcomini, A., P.D. Capel, T. Lichtensteiger, P.H. Brunner and W. Giger, 1989, Behavior of aromatic surfactants and PCBs in sludge-treated soils and landfills, *J. Environ. Qual.* 18: 523–528.

McLachlan, M.S., A.P. Sewart, J.R. Bacon and K.C. Jones, 1996, Persistence of PCDD/Fs in a sludge-amended soil, *Environ. Sci. Technol.* 30: 2567.

O'Connor, G.A., 1998, Fate and potential of xenobiotics, pp. 203–217, S. Brown, J.S. Angle and L. Jacobs (Ed.), *Beneficial Co-utilization of Agricultural, Municipal and Industrial By-products*, Kluwer Academic Publishers, Dordrecht, Netherlands.

O'Connor, G.A., D. Kiehl, G.A. Eiceman and J.A. Ryan, 1990, Plant uptake of sludge–borne PCBs, *J. Environ. Qual.* 19: 113–118.

Overcash, M.R., 1983, Land treatment of municipal effluent and sludge: Specific organic compounds, Workshop on Utilization of Municipal Wastewater and Sludge on Land, U.S. Environmental Protection Agency, U.S. Army Corps of Engineers, USDA Cooperative State Research Service, National Science Foundation, University of California–Kearney Foundation of Soil Science, Denver, CO.

Overcash, M.R., J.B. Weber and W. Tucker, 1986, Toxic and priority organics in municipal sludge land treatment systems, Rep. No. USEPA Grant No. CR806421, U.S. Environmental Protection Agency, Cincinnati, OH.

Potter, C.L., J.A. Glaser, L.W. Chang, J.R. Meier, M.A. Dosani and R.F. Herrmann, 1999, Degradation of polynuclear aromatic hydrocarbons under bench-scale compost conditions, *Environ. Sci. Technol.* 33(10): 1717–1725.

Rose, W.W. and W.A. Mercer, 1968, Fate of insecticides in composted agricultural wastes, National Canners Association, Washington, D.C.

Ryan, J.A., R.M. Bell, J.M. Davidson and G.A. O'Conner, 1988, Plant uptake of nonionic organic chemicals from soils, *Chemosphere* 17: 2299–2323.

Saber, D.L. and R.L. Crawford, 1985, Isolation and characterization of *Flavobacterium* strains that degrade pentachlorophenol, *Appl. Environ. Microbiol.* 50: 1512–1518.

Sawhney, B.L. and K. Brown, 1989, *Reactions and Movement of Organic Chemicals in Soils*, SSSA Special Pub. No. 2, Soil Science Society of America, American Society of Agronomy, Madison, WI.

Simonich, S.L. and R.A. Hites, 1995, Organic pollutant accumulation in vegetation, *Environ. Sci. Technol.* 29(12): 2905–2914.

Smith, S.R., 1996, *Agricultural Recycling of Sewage Sludge and the Environment*, CAB International, Wallingford, U.K.

Snell Environmental Group, I., 1982, Rate of biodegradation of toxic organic compounds while in contact with organics which are actively composting, National Science Foundation, PB84–193150, Washington, D.C.

Strek, H.J. and J.B. Weber, 1980, Adsorption and translocation of polychlorinated biphenyls (PCBs) by weeds, *Proc. Sci. Soc.* 33: 226–232.

Stegmann, R., S. Lotter and J. Heerenklage, 1991, Biological treatment of oil-contaminated soils in bioreactors, pp. 118–208, R.E. Hinchee and R. F. Olfenbuttel (Eds.), *On–Site Bioreclamation*, Battelle, Columbus, OH.

USEPA, 1985, Health Assessment Document for Chlorinated Benzenes, U.S. Environmental Protection Agency, Office of Research and Development, Office of Environmental Assessment, EPA-600/8-84-015F, Washington, D.C.

USEPA, 1985, Summary of Environmental Profiles and Hazard Indices for Constituents of Municipal Sludge, Office of Water Regulations and Standards, Wastewater Solids Criteria Branch, Washington, D.C.

USEPA, 1990, National sewage sludge survey: Availability of information and data and anticipated impacts on proposed regulations, U.S. Environmental Protection Agency, *Fed. Reg.* 55: 47210–47283, Washington, D.C.

USEPA, 1999, Dioxin Fact Sheet, U.S. Environmental Protection Agency, Washington, D.C.

Wang, M. and K.C. Jones, 1994, Uptake of chlorobenzenes by carrots from spiked and sewage-sludge-amended soil, *Environ. Sci. Technol.* 28(7): 1260–1267.

Webber, M.D. and S. Lesage, 1989, Organic contaminants in Canadian municipal sludges, *Waste Manage. Res.* 7: 63–82.

Welsch-Pausch, K., M.S. McLachlan and G. Umlauf, 1995, Determination of the principal pathways of polychlorinated dibenzo-*p*-dioxins and dibenzofurans to *Lolium multiflorum* (Welsh ray grass), *Environ. Sci. Technol.* 29(4): 1090–1099.

Wild, S.R. and K.C. Jones, 1992, Polynuclear aromatic hydrocarbon uptake by carrots grown in sludge–amended soil, *J. Environ. Qual.* 21: 217–225.

Wild, S.R., S.K. Waterhouse, S.P. McGrath and K.C. Jones, 1990, Organic contaminants in an agricultural soil with a known history of sewage sludge amendments: Polynuclear aromatic hydrocarbons, *Environ. Sci. Technol.* 24: 1706–1712.

Wild, S.R., J.P. Obbard, C.I. Munn, M.L. Berrow and K.C. Jones, 1991, The long-term persistence of polynuclear aromatic hydrocarbons (PAHs) in an agricultural soil amended with metal–contaminated sewage sludges, *Sci. Total Environ.* 101: 235–253.

Wild, S.R., S.J. Harrad and K.C. Jones, 1994, The influence of sewage sludge applications to agricultural land on human exposure to polychlorinated dibenzo-p-dioxins and -furans (PCDFs), *Environ. Pollut.* 83: 357–369.

Willett, L.B., 1975, Excretory behavior of polychlorinated biphenyls in lactating cows fed normal and thyroprotein containing rations, *J. Dairy Sci.* 58: 765.

Wilson, S.C., R.E. Alocock, A.P. Sewart and K.C. Jones, 1997, Persistence of organic contaminants in sewage sludge-amended soil: A field experiment, *J. Environ. Qual.* 26: 1467–1477.

Wilson, S.C. and K.C. Jones, 1999, Volatile organic compound losses from sewage sludge-amended soils, *J. Environ. Qual.* 28: 1145–1153.

CHAPTER 7

Pathogens in Wastewater and Biosolids

INTRODUCTION

A human pathogen is any virus, microorganism, or substance capable of causing disease (*Stedman's Medical Dictionary,* 1977). By this definition, bacteria, parasites, viruses, microbial substances (endotoxins), fungi and other organisms are pathogens. The two general categories are primary and secondary pathogens. Primary pathogens, such as bacteria, parasites and viruses, can invade and infect healthy humans (Burge and Millner, 1980). Secondary pathogens invade and infect a debilitated or an immunosuppressed individual. Often secondary pathogens, such as fungi, are termed *opportunistic pathogens*, since they infect those who have suffered disease, causing severe debilitation.

Fecal coliform, an indicator organism when present in large numbers, indicates the potential presence of pathogens. Intestinal pathogenic bacteria normally react to environmental conditions in a similar manner, as do coliforms. Thus, fecal coliforms are good indicator organisms.

Yanko (1988) demonstrated a strong correlation between fecal coliform densities and frequency of salmonella detection. The data showed that when the log fecal coliform density was below 3 (1000 MPN/g total solids), the frequency of detection of salmonellae was in the range of 0 to 3 MPN/g total solids. Yanko sampled biosolids compost and did not find salmonella in 86 measurements for which the fecal coliform densities were less than 1,000 MPN per gram. This was the basis for the pathogen regulations for Class A biosolids (Farrell, 1992).

One of the greatest concerns with land application of biosolids is the presence of pathogens, for the following reasons:

- Uptake by plants and entry into the food chain
- Movement through the soil and contamination of groundwater with potential contamination of drinking water
- Runoff and erosion containing pathogens and contaminating surface water. This could result in direct exposure to persons contacting the contaminated water (i.e., bathers) or through contamination of drinking water supplies.

The mere presence of a pathogen is not indicative of the potential for infection or disease. In addition to the presence of organisms, it is important to know how many organisms will cause an infection. This is called the infective dose or dose–response relationship.

Wastewater contains pathogens from human and animal wastes discharged into the sewer system. In addition, surface runoff combined with the sewer system will contain mammalian (especially animal) and avian pathogens. Global and regional conditions such as climate can also affect the type and numbers of certain pathogens. The mobility of our society, ease of travel and influx of individuals from developing countries, especially from semitropical or tropical regions, increase the likelihood of both numbers and types of parasites into wastewaters. The recent increase and appearance of several human and animal organisms or toxins such as *E. coli* 0157:H7, HIV, *Helicobacter pylori* and mad cow (bovine spongiform encephalopathy or BSE) disease could result in these organisms or toxins appearing in wastewaters and biosolids. Little is known of the effect of the wastewater treatment process on these.

E. coli 0157:H7 can produce verotoxins causing hemorrhagic colitis (diarrhea that becomes profuse and bloody), hemolytic uremic symptom (bloody diarrhea followed by renal failure) and thrombocytopenic purpura, with symptoms similar to those of hemolytic uremia that also involve the central nervous system (Pell, 1997). Outbreaks from contaminated food and water have been reported (Besser et al., 1993; Wang et al., 1996). Little data exist on virulent strains (e.g., *E. coli* 0157:H7) in biosolids.

Human immunodeficiency virus (HIV) consists of a nucleic acid core, or genome, surrounded by a shell of proteins termed *capsid*. The capsid consists of a bilipid layer, an exterior glycoprotein and a transmembrane glycoprotein. Johnson et al. (1994) discussed the implication of HIV to the wastewater industry. Their main concern was the health implication to workers. As they reported, the discharge of fluids containing the virus would be in small volumes compared to the total discharge of influents, resulting in dilution. Furthermore, the concentration of HIV in human body fluids is low in comparison to other pathogens. Once outside the human body, viable HIV concentrations decline at a first-order rate. Also, the organism cannot survive or reproduce without a host cell.

Once HIV leaves the protective environment of the host cell, it is most susceptible to deactivation and cannot reproduce. Danger to workers would be greater from handling contaminated objects such as condoms or blood-stained cotton gauzes, bandages, or sanitary napkins. The use of protective clothing is recommended.

Several authors studied the survival of HIV in wastewater (Casson et al., 1992, 1997; Enriquez et al., 1993; Slade et al., 1989). The data indicate that HIV survival in wastewater is less than 50 hours. Thus, the danger to the public from the use of biosolids is probably nonexistent. Furthermore, research with enteroviruses and polioviruses has shown that viruses tend to be adsorbed on the organic fraction and deactivated (Johnson et al., 1994). Workers nevertheless need to take precautions.

Helicobacter pylori is a human gastrointestinal pathogen involving gastritis, duodenal ulcers and gastric neoplasm (Gilbert et al., 1995). A major cause of peptic ulcer disease and gastric neoplasia, the common pathogen infiltrates about 60% of the world's population (Cave, 1997). The mode of entry to the stomach is through the mouth. Infection appears to occur mostly during childhood. Fecal oral spread is a possibility, though fecal excretion has not been demonstrated (Cave, 1997). It has been very difficult to demonstrate its presence in the environment and is not presently recovered from sewage. Cave reported that changes in sanitary conditions since World War II resulted in a substantial decrease of the organism. Grubel et al. (1997) suggested that flies may pick up *H. pylori* in human wastes, particularly from untreated sewage, and deposit contaminated fly excreta on food or even directly onto the oral mucous membranes of young children.

"Mad cow" disease is not the result of a living pathogen. This disease in humans is also referred to as Creutzfeld-Jacob. The manifestation is spongy holes in the brain that is believed to be caused by prions, which are proteins that sit on the surface of brain cells. The deadly agent is a misfolded or misshapen prion. It is believed that when an abnormal prion is ingested from food, it travels to the brain, where in some way it subverts or changes the normal prion protein into an abnormal shape. Even if contaminated food is discharged into the wastewater stream, these proteins will likely be degraded during secondary treatment. Furthermore, in soils the proteins would be a source of organic nitrogen and transformed to inorganic nitrogen. Their large molecular structure would preclude any uptake by plants.

The primary pathogens found in wastewater and biosolids can be grouped into four major categories:

- Bacteria
- Enteric viruses
- Protozoa
- Helminths
  - Nematodes (round worms)
  - Cestodes (tapeworms)

Examples of secondary pathogens in biosolids include:

- *Escherichia coli (E. coli)*
- *Klebsiella* sp.
- *Yersinia* sp.
- *Aspergillus fumigatus*
- *Listeria*

Although *E. coli* is often termed a secondary pathogen, pathogenic strains of *E. coli* can cause diarrhea and gastroenteritis (Sack, 1975). Fatalities have occurred in children. A recent outbreak in Japan infected 8000 children, resulting in several deaths.

Endotoxins and organic dust are examples of pathogenic substances that may be in biosolids or biosolid-derived products. These and other organisms can be airborne or aerosolized during land application, composting, or heat drying (Sorber et al., 1984).

On February 19, 1993, USEPA promulgated regulations for the utilization and disposal of biosolids. These regulations, titled "Standards for the Use or Disposal

Table 7.1a Some Bacteria Found in Wastewater, Sludge and Biosolids and the Diseases They Transmit

Bacteria	Disease
Salmonella spp. (approximately 1700 types)	Salmonellosis
	Gastroenteritis
Salmonella typhi	Typhoid fever
Mycobacterium tuberculosis	Tuberculosis
Shigellae (4 species)	Shigellosis
	Bacterial dysentery
	Gastroenteritis
Escherichia coli (pathogenic strains)	Gastroenteritis
Yersinia spp.	Yersinosis
Campylobacter jejuni	Gastroenteritis
Vibrio cholerae	Cholera

Data sources: Epstein and Donovan, 1992; Akin et al., 1983; Ward et al., 1984; Smith and Farrell, 1996.

of Sewage Sludge; Final Rules 40 CFR Part 503," were published in the *Federal Register* Volume 38, Number 32. The rule referred to as Part 503 governs land application of biosolids, including distribution and marketing of biosolid products. The intent of the rule was to encourage beneficial use of biosolids while protecting human health and the environment.

Pathogen and vector attraction reduction (VAR) are discussed under Subpart D, Section 503.30. Two requirements of sewage sludge with respect to pathogens must be met and one of the VAR requirements must be met. Chapter 11 discusses the federal regulations as well as state and several other country regulations. Part 503 regulations do not regulate bioaerosols or secondary pathogens.

This chapter provides information on primary and secondary pathogens in biosolids and other domestic wastes; exposure, infectivity and risk; effect of wastewater treatment on removal of pathogens; and effect of biosolids treatment on destruction of pathogens. Survival in soils and plants is covered in Chapter 8.

PATHOGENS IN WASTEWATER, SLUDGE, AND BIOSOLIDS

The objective of wastewater treatment is to remove pathogens and disinfect effluent prior to discharge into water courses. The efficiency of removal varies with the different unit processes. It also depends on the organisms and their physical and biological properties. For example, many parasites survive the wastewater treatment process and accumulate in the solids fraction, termed sludge, as a result of their densities. Parasitic eggs tend to settle out in sludge at a more rapid rate than protozoan cysts (Farrell et al., 1996).

Numerous pathogens are found in wastewater and sludge (see Tables 7.1a, b, c and d). The pathogenic bacteria of major concern are *E. coli* (pathogenic strains), *Salmonella* sp., *Shigella* sp. and *Vibrio cholerae* (Kowal, 1985).

The type and densities of pathogens in biosolids are primarily a function of the wastewater and biosolids treatment processes. Pedersen et al. (1981) found that the

Table 7.1b Some Viruses Found in Wastewater, Sludge and Biosolids and the Diseases They Transmit

Virus	Disease
Adenovirus (31 types)	Conjunctivitis, respiratory infections, gastroenteritis
Polio virus	Poliomyelitis
Coxsackievirus	Aseptic meningitis, gastroenteritis
Echovirus	Aseptic meningitis
Reovirus	Respiratory infections, gastroenteritis
Norwalk agents	Epidemic gastroenteritis
Hepatitis viruses	Infectious hepatitis
Rotaviruses	Gastroenteritis, infant diarrhea

Data sources: Epstein and Donovan, 1992; Ward et al., 1984; Smith and Farrell, 1996.

primary way to reduce pathogenic organisms is by removing their food sources. The majority of the data on pathogens in biosolids, as a result of the wastewater treatment, has been generated prior to 1980.

Sekla et al. (1980) isolated 54 strains of salmonella from 38 samples of sludge and 16 samples of effluent, representing 13 serotypes. Theis et al. (1978) reported that positive samples of helminth were recovered from sludge from Los Angeles, Sacramento and Oakland, California; as well as Springfield, Missouri; Hopkinsville, Kentucky and Frankfort, Indiana.

Koenraad et al. (1997) found that the numbers of *Campylobacter* in wastewater in the United Kingdom, Germany, Italy and the Netherlands ranged from 50 to more than 50,000 MPN/100 ml. Ten species are known to infect humans, resulting in enteritis, fever, gingivitis, periodontitis and diarrhea. Cliver (1975) recovered human intestinal viruses from waste and return-activated sludge. The enteroviruses included poliovirus and reovirus.

Wellings et al. (1976) isolated Echo-7 virus from biosolids after 13 days on biosolid-drying beds. Moore et al. (1978) showed that 89% to 99% of the viruses were associated with solids from activated sludge aeration basins. In four cities that were studied, enteroviruses were detected in the range of 190 to 950 PFU/l. Grabow (1968) and Foster and Engelbrecht (1973) reported that more than 100 distinct serotypes of viruses are present in wastewater. Their data are summarized in Table 7.2.

Individuals exposed to a pathogenic organism may not necessarily become infected. The dose–response relationship is an indication of the infective dose. This dose–response is difficult to assess since tolerance for individuals varies widely (Jones et al., 1983). Furthermore, infection does not necessarily result in a disease. Table 7.3 shows dose–response for several pathogens (Bryan, 1977).

Akin (1983) reviewed the literature on infective dose data for enteroviruses and other pathogens. The widest dose response range occurred with enteric bacteria. *Salmonella* spp. required 10^5 to 10^8 cells to produce a 50% disease rate in healthy adults. Three species of *Shigella* produced illness in subjects administered 10 to 100 organisms. Administering small doses, 1 to 10, cysts of *Entamoeba coli* and *Giardia lamblia* caused amoebic infections. Very low doses of enteric viruses were found to produce infection. Hornick et al. (1970) administered various doses of *Salmonella*

Table 7.1c Some Protozoa and Helminth Parasites Found in Wastewater, Sludge and Biosolids and the Diseases They Transmit

Organism	Disease
Protozoa	
Entamoeba histolytica	Amoebic dysentery, amebiasis
Giardia lamblia	Giardiasis
Balantidium coli	Balantidiasis
Naegleria fowleri	Meningoencephalitis
Cryptosporidium spp.	Gastroenteritis
Toxoplasma gondii	Toxoplasmosis
Helminths – Nematodes	
Ascaris lumbricoides	Ascariasis
Ascaris suum	Respiratory
Ancylostoma duodenale	Hook worm, ancylostomiasis
Necator americanus	Hookworm
Ancylostoma braziliense (cat hookworm)	Cutaneous larva migrans
Ancylostoma caninum (dog hookworm)	Cutaneous larva migrans
Enterobius vermicularis (pinworm)	Enterobiasis
Strongyloides stercoarlis (threadworm)	Strongyloidiasis
Toxocara cati (cat roundworm)	Visceral larva migrans
Toxocara canis (dog roundworm)	Visceral larva migrans
Trichuris trichiura (whip worm)	Trichuriasis
Helminths – Cestodes	
Taenia saginata (Beef tapeworm)	Taeniasis
Taenia solium (pork tapeworm)	Taeniasis
Necator americanus	Hookworm disease
Hymenolepis nana (dwarf tapeworm)	Taeniasis
Echinococcus granulosus (dog tapeworm)	Unilocular echinococcosis
Echinococcus multilocularis	Alveolar hydatid disease

Data sources: Akin et al., 1983; Epstein and Donovan, 1992; Smith and Farrell, 1996.

typhi to 14 adult volunteers and found that none showed any symptoms when 1000 organisms were administered. When a dose of 100,000 organisms was administered, 28% of the adults became ill; 95% of the subjects were ill when 1,000,000,000 organisms were administered.

Table 7.1d Pathogenic Fungi that May be Present in Sludge and Biosolids

Fungi	Disease
Aspergillus fumigatus	Respiratory infections
Candida ablicans	Candidiasis
Cryptococcus neoformans	Subacute chronic meningitis
Epidermophyton spp. and Trichophyton spp.	Ringworm and athlete's foot
Trichosporon spp.	Infection of hair follicles
Phialophora spp.	Deep tissue infections

Source: Adapted from Fradkin, 1989.

Table 7.2 Viruses in Wastewater and Sewage Sludge

Virus	Disease
Hepatitis A virus	Infectious hepatitis
Norwalk and Norwalk-like viruses	Gastroenteritis
Rotaviruses	Gastroenteritis
Enteroviruses	
Poliovirus	Poliomyelitis
Coxsackieviruses	Meningitis, pneumonia, hepatitis, cold-like symptoms
Echoviruses	Meningitis, encephalitis,cold-like symptoms
Reovirus	Respiratory infections, gastroenteritis
Astroviruses	Gastroenteritis
Caliciviruses	Gastroenteritis

Source: USEPA, 1999.

Table 7.3 Dose-Response for Several Pathogens

Pathogen	Approximate Dose to Produce Disease in 25-75% of Subjects Tested	Minimum Dosage to Produce Disease in Any Individual Number of Organisms
Shigella sp.	10^2–10^5	10^1
Salmonella sp.	10^5–10^9	10^4
Escherichia coli	10^6–10^{10}	10^6
Vibrio cholerae	10^3–10^{11}	10^3
Streptococcus faecalis	$>10^{10}$	10^{10}
Entamoeba coli[1]	10^1–10^3	10^1
Giardia lamblia[1]	–	10^1

[1] The dosage caused infection and not the disease.

Source: Adapted from Bryan, 1977.

Pharen (1987) reviewed the literature on infective doses for bacteria and viruses. In addition to the infective dose, other factors, such as age and general health, are important. Pharen states, "However, people do not live in a germ- nor risk-free society. Microorganisms are present almost everywhere — in the air, the soil and on objects that people touch." Additional information on the infective dose data as reported in the literature is shown in Table 7.4.

Table 7.4   Reported Infective Dose for Several Organisms

Organism	Infective Dose	Range	Reference
	Bacteria		
Clostridium perfringens	10^6	10^6–10^{10}	Kowal, 1985
Escherichia coli	10^4	10^4–10^{10}	Keswick, 1984; Kowal, 1985
Salmonella (various species)	10^2	10^2–10^{10}	Kowal, 1985;
Shigella dysenteriae	10–10^2	10–10^9	Kowal, 1985; Keswick, 1984; Levine et al., 1973
Shigella flexneri	10^2	10^2–10^9	Kowal, 1985
Streptococcus faecalis	10^9	10^9–10^{10}	Kowal, 1985
Vibro cholerae	10^3	10^3–10^{11}	Kowal, 1985; Keswick, 1984
	Viruses		
Echovirus 12	HID_{50}[a] 919 PFU[b] $HID1$[c] 17 PFU estimated	17–919 PFU	Kowal, 1985
Polio virus	1 $TCID_{50}$[d] , <1 PFU	$4 \times 10^7 TCID_{50}$ for infants; 0.2–5.5×10^6 PFU for infants	Kowal, 1985
Rotavirus	HID_{50} 10 ffu HID_{25} 1 ffu estimated	0.9–9×10^4	Ward et al., 1986
	Parasites		
Entamoeba coli	1–10 cysts	1–10 cysts	Kowal, 1985
Cryptosporidium	10 cysts 30 oocysts	10–100 cysts	Casmore, 1991 Dupont et al., 1995
Giardia lamblia	1 cyst estimated	NR	Kowal, 1985
Helminths	1 egg	NR	Kowal, 1985

[a] HID = Human infective dose.
[b] Plaque forming units per gram dry weight.
[c] $TCID_{50}$ = 50% tissue culture infectious dose.
[d] ffu = focus forming units.

REMOVAL OF PATHOGENS BY WASTEWATER TREATMENT PROCESSES

Several physical, chemical and biological factors inactivate pathogens. Reimers et al. (1996) discuss these factors, which are summarized in Table 7.5.

The type and densities of pathogens in biosolids is primarily a function of the wastewater and biosolids treatment processes. Pedersen et al. (1981) indicate that the primary reduction of pathogenic organisms results through removal of the food sources. The majority of the data on pathogens in biosolids undergoing wastewater

Table 7.5 Physical, Chemical and Biological Factors Affecting Inactivation of Pathogens

Physical	Chemical	Biological
Temperature	pH (acids/alkali)	Antagonistic organisms
Applied fields	Ozone	Digestion (aerobic/anaerobic)
Microwave irradiation	Ammonia	Composting
Infrared irradiation	Nitrous acids	Alkaline composting
Ultra sonication	Phosphoric acid	
Magnetic fields	Nitric acid	
Pulsing electrostatic/electrolytics	Alkaline agents	
Desiccation	Sulfuric acid	

Source: Reimers et al., 1996, pp. 51–74, Stabilization and Disinfection — What Are Our Concerns, Water Environment Federation, Dallas, TX. With permission.

treatment has been generated prior to 1980. Table 7.6 provides some of the early data (Pedersen et al., 1981). Data on viruses were limited due to poor recovery from solids. Although methodologies for the enumeration of pathogens in biosolids have been shown to be deficient, updates in more recent years have been scant (Yanko et al., 1995). Parsons et al. (1975) summarized findings in the literature at that time on the effect of wastewater treatment on pathogen destruction. The authors concluded that wastewater treatments significantly reduced certain pathogenic microorganisms, but no single process yielded an effluent virtually free of pathogenic microorganisms.

During primary and secondary treatment, many pathogens are destroyed. Foster and Engelbrecht (1973) summarized the early data, shown in Table 7.7. Many of the pathogens removed during primary and secondary treatment will be associated with the biosolids. Land application of biosolids requires disinfection and stabilization. Dahab et al. (1996) determined the concentrations of fecal coliform, fecal streptococci and *Salmonella* spp. in primary sludge in nine different wastewater treatment plants. Fecal coliform densities varied from 12 to 61 million MPN/g of total solids (TS), the most probable number per gram of total solids. The average was 36 million MPN/g of TS. Fecal streptococcus densities ranged from a low of 2.6 million to a high of 40 million MPN/g TS. *Salmonella* spp. densities varied from 217 to 1000 MPN/g TS for eight of the treatment plants. At the ninth plant, the levels were 3140 MPN/g of TS.

Stadterman et al. (1995) evaluated the efficiency of the removal of *Cryptosporidium* oocysts by the waste-activated sludge treatment and anaerobic digestion. The authors reported that the total oocyst removal in sewage treatment was 98.6%. After 24 hours 99.9% of the oocysts were eliminated by anaerobic digestion. Koenraad et al. (1997) found that the wastewater treatment processes reduced the levels of *Campylobacter* by several factors, but many of the organisms survived. Anaerobic digestion had little effect on reducing the numbers, but aerobic digestion was effective in eliminating the organism.

Malina (1976) reported an early review of the inactivation of viruses by various wastewater treatment processes. Some of the data is summarized in Table 7.8. The author points out that in many of the studies, the virus titer was far in excess of

Table 7.6 Density Levels of Indicator Organisms and Pathogens in Primary, Secondary and Mixed Biosolids[a]

Organism	Primary		Secondary		Mixed	
Total coliform bacteria	1.2×10^8	Gaby, 1975; Noland et al., 1978	7.1×10^8	Noland et al., 1978; Bovay Engineers,1975	1.1×10^9	Berg & Berman, 1980; Laconde et al., 1978a; b
Fecal coliform bacteria	2.0×10^7	Gaby, 1975; Noland et al., 1978; Counts & Shuckrow, 1974; SAC, 1979	8.3×10^6	Noland et al., 1978; Bovay Engineers, 1975; Counts & Shuckrow, 1974	1.9×10^5	Counts & Shuckrow, 1974; Berg & Berman, 1980; Laconde et al., 1978a; b
Fecal streptococci	8.9×10^5	Gaby, 1975; Noland et al., 1978; Counts & Shuckrow, 1974; SAC, 1979	1.7×10^6	Noland et al., 1978; Bovay Engineers, 1975; Counts & Shuckrow, 1974	3.7×10^6	Counts & Shuckrow, 1974; Berg & Berman, 1980; Laconde et al., 1978a; b
Salmonella sp.	4.1×10^2	Noland et al. 1978; Counts & Shuckrow, 1974; SAC, 1979; Moore et al., 1978	8.8×10^2	Noland et al., 1978; Counts & Shuckrow, 1974	2.9×10^2	Counts & Shuckrow, 1974; Laconde et al., 1978a, b
Pseudomonas aeruginosa	2.8×10^3	Noland et al., 1978; Counts & Shuckrow, 1974	1.1×10^4	Noland et al., 1978; Counts & Shuckrow, 1974	3.3×10^3	Counts & Shuckrow, 1974
Ascaris sp.	7.2×10^2	Reimers et al., 1980	1.4×10^3	Reimers et al., 1980	2.9×10^2	Reimers et al., 1980
Trichuris trichiura	1.0×10^1	Reimers et al., 1980	$<1.0 \times 10^1$	Reimers et al., 1980	0	Reimers et al., 1980
Trichuris vulpis	1.1×10^2	Reimers et al., 1980	$<1.0 \times 10^1$	Reimers et al., 1980	1.4×10^2	Reimers et al., 1980
Toxocara sp.	2.4×10^2	Reimers et al., 1980	2.8×10^2	Reimers et al., 1980	1.3×10^3	Reimers et al., 1980
Hymenolpepis diminuta	6.0×10^0	Reimers et al., 1980	2.0×10^1	Reimers et al., 1980	0	Reimers et al., 1980
Enteric viruses[b]	3.9×10^2	Nath & Johnston, 1979; Moore et al., 1978; Hurst et al., 1978; Nielsen & Lydholm, 1980	3.2×10^2	Moore et al., 1978; Hurst et al., 1978; Nielsen & Lydholm, 1980	3.6×10^{2c}	Nielsen & Lydholm, 1980

[a] Data are average geometric means of organisms per gram solids dry weight.
[b] Plaque forming units per gram dry weight (PFU/gdw).
[c] $TCID_{50}$ = 50 percent tissue culture infectious dose.

Source: Pedersen, 1981.

Table 7.7 Pathogen Removal Efficiency during Primary and Secondary Wastewater Treatment

Pathogen	Primary Treatment % Removal Efficiency	Secondary Treatment % Removal Efficiency	Trickling Filter % Removal Efficiency
Bacteria[a]	50–90	90–99	90–95
Salmonella sp.[b]	15	96–99	84–99.9
Mycobacterium sp.[b]	48–57	Slight to 87	66–99
Protozoan cysts[a]	10–50	50	50–95
Amoebic cysts[b]	No reduction in 3 hours	No apparent removal	11–99.9
Helminth ova[a]	72–78	No apparent removal	62–76[b]
			50–90[a]
Virus[ab]	3 to extensive removal[b]	76–99[b]	0–84[b]
	0–30[a]	90–99[a]	90–95[a]

[a] Feachem et al. (1980)
[b] Foster and Engelbrecht (1973)

indigenous levels of 4000 to 7000. However, inactivation would occur also at lower levels. The data show that wastewater treatment is only partially effective in the inactivation of viruses but, chlorine or ozone disinfection is very effective. Removal of organisms during wastewater treatment displaces them from the liquid stream. However, they become associated with the solids.

EFFECT OF BIOSOLIDS TREATMENT

The solids resulting from wastewater treatment must undergo further treatment prior to land application. Land application of biosolids requires the disinfection and stabilization of biosolids. The objective is to reduce the level of pathogens, reduce vector attraction and produce a stabilized product — that is, a product that would not decompose very rapidly and produce offensive odors. Table 7.9 shows the general effect of wastewater treatment and densities of microorganisms in effluent and biosolids (NRC, 1996).

Temperature is very effective in the destruction of pathogens. The time–temperature relationships for pathogen destruction were used in the USEPA 503 regulations (USEPA,1992). Several biosolid processes rely on temperature to meet Class A biosolids. These include: composting, heat drying, alkaline stabilization and thermal digestion. Table 7.10 shows the thermal destruction of several pathogens and parasites. This data is derived from pathogen destruction in liquids where temperature is much more uniform throughout the mass. With biosolid and biosolid products, a longer period of time is needed to ensure that every particle within the mass is subjected to the temperature.

USEPA in the 503 regulations requires that either a Class A or B biosolid be produced prior to land application. Class A biosolid is a material that has undergone treatments that reduce pathogens to very low or undetectable levels. A less stringent requirement is allowed for Class B. Details of the regulations are provided in Chapter 11.

Table 7.8 Effect of Municipal Wastewater Treatment on Viral Inactivation

Wastewater Process	Virus	Percent Removal	Titer PFU/l	Reference
Primary clarification	Bacteriophage F2	37.1	$(6.7\text{–}7.6) \times 10^5$	Sherman, 1975
	Polio 1 (Mahoney)	26–55	2×10^8	Clarke et al., 1964
	Polio 1,2,3 (Sabin)	0–12	*	England et al., 1967
Activated biosolids	Bacteriophage T2	98	$(3\text{-}50) \times 10^5$	Kelly et al., 1961
	Coxsackie A9	96–99.4	3×10^8	Clarke et al., 1961
	Polio 1 (Sabin)	98	7.7×10^4	Malina and Melbard, 1974
	Polio 1 (MK 500)	64–78	$(2\text{–}200) \times 10^6$	Kelly et al., 1961
	Polio 1 (Mahoney)	79–94	7×10^7	Clarke et al., 1961
	Polio 1,2,3 (Sabin)	76–90	*	England et al., 1967
Aerated lagoons	Polio 1 (Sabin)	99	1.6×10^3	Ranganathan et al., 1974
Oxidation ponds	Polio 1 (attenuated)	92	5.6×10^5	Malina et al., 1975
	Polio 1 (Sabin)	99	3.3×10^3	Malina and Melbard, 1974
	Reovirus	95	20,000*	Nupem et al., 1974
	Polio 1 (Mahoney)	99.97-ND	$(6\text{–}1800) \times 10^3$	Malina and Melbard, 1974
Trickling filter	Bacteriophage F2	18.9	$(5.9\text{–}7.5) \times 10^5$	Sherman, 1975
	Coxsackie A9	84	3×10^9	Clarke and Chang, 1975
	Echovirus 12	83	7×10^9	Clarke and Chang, 1975
	Polio 1	85	4×10^9	Clarke and Chang, 1975
Disinfection chlorine	Bacteriophage F2	99.997	No data	Anonymous, 1975
	Polio	99	10^{9**}	Kott et al., 1975
Disinfection ozone	Bacteriophage F2	100	1×10^{11}	Pavoni and Tittlebaum, 1974
	Cocsackie B3	99.9	2.5×10^2	Keller, 1974
	Polio I	99.994	$(1.4\text{–}6.3) \times 10^7$	Majumdar et al., 1974
	Polio II	99.99	5×10^2	Keller, 1974

*Natural levels following immunization.
**$TCID_{50}/l$ = 50% tissue culture infectious dose.
Source: Adapted from Malina, 1976.

The processes used to achieve the regulatory requirements are:

- Aerobic digestion
- Anaerobic digestion
- Composting
- Heat drying
- Alkaline stabilization

Table 7.9 Effect of Wastewater Treatment on the Densities of Microorganisms in Effluent and Biosolids

Organism	Raw Sewage	Number per 100 ml of Effluent			Numbers per gram of Biosolids	
		Primary Treatment	Secondary Treatment	Tertiary[a] Treatment	Raw	Digested[b]
Fecal coliform MPN[c]	1×10^9	1×10^7	1×10^6	<2	1×10^7	1×10^6
Salmonella MPN	8,000	800	8	<2	1,800	18
Shigella MPN	1,000	100	1	<2	220	3
Enteric virus PFU[d]	50,000	15,000	1,500	0.002	1,400	210
Helminth ova	800	80	0.08	<0.08	30	10
Giardia lamblia cysts	10,000	5,000	2,500	3	140	43

[a] Includes coagulation, sedimentation, filtration and disinfection.
[b] Mesophilic anaerobic digestion.
[c] MPN = most probable number.
[d] PFU = plaque forming units.
References: USEPA, 1991 and 1992; Dean and Smith, 1973; Feachem et al., 1980; Engineering Science, 1987; Gerba, 1983.

Table 7.10 Thermal Destruction of Several Pathogens and Parasites

Organism	Thermal Death Points
Salmonella typhosa	No growth beyond 46°C; death within 30 min at 55° to 60°C
Salmonella spp.	Death within 1 h at 56°C; death within 15 to 20 min at 60°C
Shigella spp.	Death within 1 h
Escherichia coli	Most die within 1 h at 55°C and within 15 to 20 min at 60°C
Micrococcus pyogenes var. *aureus*	Death within 10 min at 50°C
Streptococcus pyogenes	Death within 10 min at 54°C
Microbacterium tuberculosis var. *hominis*	Death within 15 to 20 min at 66°C
Mycobacterium diptheriae	Death within 45 min at 55°C
Endamoeba histolytica	Thermal death is 68°C
Tania saginata	Death within 5 min at 71°C
Trichinella spiralis larvae	Thermal death point is 62 to 72°C
Necator americanus	Death within 50 min at 45°C

Source: NRC, 1996.

Aerobic Digestion

Aerobic digestion has been carried out under mesophilic conditions ranging in temperature from ambient to 37°C and retention times of 10 to 20 days. More recently, there has been an evaluation of thermophilic aerobic digestion in order to meet Class A biosolids. Relatively few data exist on the effectiveness of aerobic digestion on pathogen destruction. Novak et al. (1984) studied the survival of indicator organisms during mesophilic aerobic digestion of biosolids at various sludge ages in the oxidation ditch. The data were variable. One-log reduction in Fecal streptococci occurred in 1 to 16 days at 20°C. For a 2-log reduction of the organisms at 20°C, an aeration time ranging from 6 to 40-plus days was required. Kebina and Plosheva (1974), as cited by Fitzgerald and Ashley (1977), found that *Ascaris suum* ova in mesophilic aerobically stabilized biosolids failed to develop in the absence of oxygen. The ova were able to develop after aeration at 20 to 27°C. *Salmonella* sp. and *E.Coli* were reduced in density by 1 log after several days of retention. Farrah et al. (1981) reported that aerobically digested biosolids contained enteric viruses from 14 to 260 $TCID_{50}/g$ (50% tissue culture infectious dose).

Autothermal thermophilic aerobic digestion (ATAD) is a relatively new process designed to produce PFRP or Class A Biosolids for land application. Therefore, data on pathogen destruction are limited. Vik and Kirk (1996) reported that since the inception of the first ATAD in 1994, fecal coliform levels measured weekly were less than 244 MPN/g of TS and most of the levels were below 100 MPN/g of TS.

Dahab et al. (1996) determined densities of fecal coliform, fecal streptococcus and *Salmonella* sp. in four wastewater treatment plants. The range of fecal coliform densities ranged from 50,000 to 3.8 million MPN/g of TS with an average density of 1.7 million MPN/g of TS. One of the four plants could not meet the USEPA Class B biosolids criteria. This was believed by the authors to be as a result of relatively low hydraulic retention time and low sludge age. Fecal streptococcus densities ranged from 30,000 to 2.23 MPN/g of TS. The average fecal streptococcus density was 850,000 MPN/g of TS. *Salmonella* sp. varied considerably. Two of the four plants had densities of 80 and 82 MPN/4g of TS and two plants had densities of 2340 and 3840 MPN/4g of TS. *Cryptosporidium* oocysts were destroyed during thermophilic aerobic digestion when temperatures exceeded 55°C (Whitmore and Robertson, 1995).

In many cases, aerobically digested biosolids can meet USEPA Class B biosolids and be land applied. ATAD systems may be able to meet Class A biosolids.

Anaerobic digestion

Anaerobic digestion can be carried out under mesophilic or thermophilic conditions. Mesophilic anaerobic digestion is usually achieved at temperatures 30°C to 38°C, whereas in thermophilic anaerobic digestion the temperatures range from 50°C to 60°C. Primary or secondary sludge is fed intermittently or continuously into sealed vessels that preclude free oxygen. Although the primary purpose of anaerobic digestion is solids reduction, other benefits are methane production and pathogen reduction.

Table 7.11 Density Levels and Reduction of Indicator Bacteria and Pathogenic Microorganisms by High Rate Mesophilic Anaerobic Digestion

Organism	Density Level[1] per 100 ml	Log Reduction Mean[2]	Log Reduction Range	Reference
Total coliform	3×10^7	2.05	1.78–2.30	Berg and Berman, 1980; Lue-Hing et al., 1979; Jewell et al., 1980
Fecal coliform	2×10^6	1.84	1.44–2.30	SAC 1979; Berg and Berman, 1980; Lue-Hing et al., 1979; Jewell et al., 1980
Fecal streptococcus	9×10^5	1.48	1.10–1.94	SAC, 1979; Berg and Berman, 1980; Lue-Hing et al., 1979; Jewell et al., 1980
Salmonella sp.	3.7×10^1	1.63	0.92–2.08	Berg and Berman, 1980; Lue-Hing et al., 1979; Jewell et al., 1980
Pseudomonas aeruginosa	6×10^5	0.58	0.15–1.36	Berg and Berman, 1980; Lue-Hing et al., 1979; Jewell et al., 1980
Enterovirus	7.9×10^1	1.21	1.05–1.36	Berg and Berman, 1980; Jewell et al., 1980

Source: Pedersen et al., 1981.

Proper mesophilic anaerobic digestion results in biosolids' meeting PSRP classification requirements or having a density of $<2 \times 10^6$ fecal coliform bacteria.

Salmonella sp. can survive mesophilic anaerobic digestion. Jones et al. (1983) reported that *Salmonella* sp. sampled during 12 days at two different treatment plants ranged from 3 to 24,000/100 ml in raw sewage and from 3 to 350/100 ml. On most days, there occurred several log reductions in organisms during 2 days in one plant. On 1 day in the second plant, the number of organisms was higher in the digested biosolids than in the raw sewage.

High rate mesophilic anaerobic digestion generally reduced pathogenic organisms and indicator bacteria by 1 to 2 logs as shown by a review of the literature prior to 1981 (see Table 7.11, Pedersen et al., 1981). New York City conducted one of the more comprehensive evaluations of pathogens and indicator organisms. Table 7.12 shows levels of pathogens and indicator organisms in anaerobically digested biosolids (NYCDEP, 1992). The biosolids were anaerobically digested for 20 days at 35°C. The data from New York showed that many pathogens and indicator organisms survived anaerobic digestion.

Soares et al. (1994) monitored enteroviruses and *Giardia* cysts in mesophilic anaerobically digested biosolids over a 14-month period. Enteroviruses ranged from 4.36×10^3 to 7.00×10^5 MPN/kg before anaerobic digestion and from 6.25 to 2.52 $\times 10^5$ MPN/kg after.

A study at three small wastewater treatment plants in British Columbia evaluated the efficiency of ATAD. At 60 to 70°C and a hydraulic retention time of about 10 days, fecal coliform and fecal streptococcus, in seven of 12 samplings, were reduced

Table 7.12 Pathogens and Indicator Organisms in Mesophilic Anaerobically
         Digested Biosolids*

Organism	Round 1 Sampling No./100 ml	Round 2 Sampling No./100 ml
Total coliform (plate count)	110,000–9,900,000	14,000–19,000,000
Fecal coliform (plate count)	11,000–620,000	1,100–4,800,000
Fecal coliform (plate count)	1,100–350,000	1,100–2,400,000
Fecal streptococci	1,100–650,000	1,100–1,300,000
Enterococci (membrane filtration)	2,100–590,000	1,100–1,200,000
Clostridium perfringens	1,100–8,500,000	2,100,000–34,000,000
Salmonella	0.08–30	0.061–3
E. coli C.	5,000–890,000	500–30,000,000
Giardia lamblia	0.0–120	20–80
Ascaris lumbricoides	0.0–67	25–100
Ancylostoma necator	0.0–33	Not recovered
Enterobius vermicularis	0.0–33	Not recovered
Trichuris trichiura	0.0–33	25–100
Total parasites	13–170	6.2–280

Source: NYCDEP, 1992.

to less than 100 MPN/gram. Five samples had fecal streptococcus at or above 100 MPN/g. No salmonellae were detected (Kelly, 1991).

Cram (1943) found viable *Ascaris* eggs after 6 months of anaerobic digestion at mesophilic temperatures of 20–30 °C. Early studies at the University of Illinois showed that the ova of the round worm, *Ascaris l. suum*, survived the anaerobic digestion process and subsequently embryonated. The sludge digestion process protected the ova from temperatures of 38°C. When the ova were subsequently exposed to air, they embryonated. Stern and Farrell (1977) indicated that *Ascaris* ova survived thermophilic digestion at 50°C. Stadterman et al. (1995) reported that anaerobic digestion at 37°C inactivated 50% of *Cryptosporidium oocysts* after 2 hours of digestion, and after 24 h 99.98% were inactivated. However, Whitmore and Robertson (1995) indicated that mesophilic anaerobic digestion at temperatures ranging from 35°C to 37°C is incapable of destroying all oocysts.

Table 7.13 shows data on enteric virus concentration in anaerobically digested biosolids. Viruses ranged from <0.03 to 210 PFU.

Table 7.13 Inactivation of Poliovirus in
         Composted Biosolids at 60%
         Moisture

Treatment	Percentage Recovery of Plaque-Forming Units
35°C, 200 min	30
39°C, 20 min	7.2
43°C, 20 min	0.087
47°C, 5 min	0.003

Composting

The effect of composting on pathogen destruction is discussed in detail in the book *The Science of Composting* (Epstein, 1997). The composting process is capable of disinfecting wastes. However, either due to poor design or poor operations, the composting process enables some pathogens to survive (Yanko, 1988). The destruction of poliovirus in 40% compost solids, as related to temperature and treatment time, is shown in Table 7.13. At 35°C, poliovirus survived for a much longer period than at 47°C. Data on heat inactivation of total coliforms, fecal coliforms, fecal streptococcus and *Salmonella enteritidis* serotype Montevideo also showed substantial reduction of organisms at temperatures of 55°C to 65°C.

Knoll (1961) also described several experiments where he subjected different salmonella strains to composting temperatures at the Baden-Baden Biosolids-refuse composting plant. After 14 days of reactor time with temperatures of 55 °C to 60°C and a moisture content of 40% to 60%, the product did not contain pathogens.

Wiley (1962) reviewed some of the early literature on pathogen destruction by composting. He reported that pathogen destruction during composting is the result of thermal kill and antibiotic action, or by the decomposing organisms or their products. Knoll (1961) tested the theory that, besides temperature, antibiotic substances resulted in pathogen destruction by composting. He extracted a solution from composted material at different stages of the process. In a compost extract taken between days 7 and 16, no inoculated *Salmonella cairo* were able to grow. He determined that the development of inhibitors originated in the presence of actinomycetes and molds and concluded that this phenomena is due to an unknown antibiotic-producing organism. Although the optimum temperature that produces the antibiotic was not determined, he postulated that 50°C to 55°C appeared to be the temperature at which these substances were generated. Gaby (1975) did not find any oppressive or antagonistic material in compost. Golueke (1983) pointed out that indigenous organisms are in a better position to compete for nutrients than pathogenic microorganisms. Furthermore, he indicated that time acts as a factor since thermal destruction is not instantaneous. Time provides for the combination of several inhibitory factors to act on pathogenic organisms.

In 1969, Morgan and Macdonald investigated the fate of *Mycobacterium tuberculosis* during open windrow composting of biosolids and refuse at the U.S. Public Health Service–Tennessee Valley Authority research and demonstration compost plant in Johnson City, Tennessee. They found that the organism was destroyed after 10 days of composting when the temperature averaged 60°C. They also discovered that if the temperature remained low, the bacteria survived for long periods of time (until the temperature exceeded 44°C for extensive periods). They also pointed out that equipment used for handling compost that was less than 17 days old should not be used with finished compost to prevent reintroduction of pathogens into the finished product.

Gaby (1975) reported on a series of studies on pathogen reduction during windrowing of refuse-biosolids composting. *Salmonella* and *Shigella*, either originally present or introduced into refuse-biosolids mixtures, were not found within 7 to 21 days. Enteroviruses were not found in the raw refuse, biosolids, or a mixture of the two materials. Poliovirus Type 2 was introduced into the windrows, but was inactivated after 3 to 7 days. Human parasitic cysts and ova were introduced into the center of the windrow, but they disintegrated after 7 days. Dog parasitic ova did, however, survive for 35 days. *Leptospira philadelphia*, a spirochaete, did not survive for more than two days after introduction into the windrow.

Krogstad and Gudding (1975) inoculated solid waste and biosolids with *Salmonella typhimurium*, *Serratia marcescens* and *Bacillus cereus*. Periodic measurements were made to determine the die-off rate. The organism could not be detected after 4 days when the temperature in a horizontal drum composter was maintained at around 65°C. They concluded that 3 to 5 days in a reactor vessel with temperatures of 60°C to 65°C would destroy the pathogens studied. Walke (1975) monitored *Escherichia coli*, *Salmonella eidleberg* and *Candida albicans* during windrow composting of bark-biosolid mixtures. The initial compost contained these organisms at a level of 10^6 microbes per dry gram. After 24 hours, the levels were 11, 130 and 620 microbes per dry gram of solids for *E. coli*, *Salmonella* sp. and *Candida albicans*, respectively. No organisms were detected after 36 h.

From 1973 to 1978, the U.S. Department of Agriculture conducted numerous studies on pathogen survival during composting by both the windrow and aerated static pile methods (Burge et al., 1978; Burge and Cramer, 1974; Epstein et al., 1977). The studies showed that salmonellae increased in growth initially, but was destroyed within 10 days of composting in the static pile and within 15 days in the windrow method. The destruction of an indicator virus, F2 bacteriophage, took 45 to 70 days in the windrow method and approximately 13 days in the aerated static pile method. This indicator virus was selected because it was more resistant to inactivation by heat than enteric pathogens, including viruses, bacteria, protozoa cysts and helminth ova.

Pathogenic microbial antagonism has been studied by several researchers (Brandon and Neuhauser, 1978; Millner et al., 1987). Millner et al. (1987) found that the types and numbers of different organisms affected the growth of salmonellae organisms. The presence of coliforms only or metabolically active bacteria and actinomycetes resulted in the death of salmonellae in compost.

During composting, three factors can result in the destruction of pathogens:

1. Time–temperature
2. Production of ammonia
3. Presence of competing organisms

Some concern for regrowth of pathogens exists, even when high temperatures have been achieved. Organisms could move from cooler sections of piles or windrows into areas that have previously been hot enough to eliminate pathogens. It is important to ensure uniform high temperatures in piles and windrows. Furthermore, by achieving stabilization, the food source is diminished and pathogens cannot compete with indigenous microbial populations for the remaining food. Burge et al. (1978) has shown evidence of this and also found that additional die-off of pathogens occurred during curing.

Heat Drying

Heat drying should completely destroy pathogens. Farrell and Stern (1975) rated heat treatment at 195°C as an excellent method of pathogen destruction. Heat treatment was rated poor for putrefaction potential and odor attenuation. Since antagonistic or competitive microorganisms are destroyed, contamination of the product can result in the growth of a pathogen to very high levels (Ward and Brandon, 1977; Brandon et al., 1977).

Alkaline Stabilization

Lime treatment of biosolids was recognized early as a method of deodorizing and disinfecting the material. USEPA 503 regulations require that the pH of biosolids be increased to 12.0 for a minimum of 2 hours. Because ammonia is released during the addition of lime, this compound could act as a disinfectant. If the addition of lime is insufficient to maintain the pH for the time required to disinfect the biosolids, the pH will drop and surviving bacteria will grow when conditions become favorable.

Pedersen et al. (1981) indicated that lime stabilization can result in 1 to 7 log reductions in some indicator and pathogenic organisms. Sepp (1980) concluded that helminth ova remained viable for long periods of time even when the pH of biosolids was at 12. Counts and Shuckrow (1974) found a 7-log reduction in fecal coliform at a pH of 12.4 in 2% biosolids. Lower values were obtained for 4.4% solids. For *Salmonella* spp., approximately 1 to 2 log reductions occurred. Burnham (1986) reported that, within 5 weeks, 25% and 35% cement kiln dust (CKD) and lime were effective in meeting PFRP (Class A Biosolid) and resulted in the destruction of *Salmonella*. Enterovirus levels were controlled to PFRP levels within 1 day by CKD and lime treatment. *Ascaris* egg survival was reduced by more than 3 log at high CKD and lime treatment within 4 weeks.

CONCLUSION

The data show that wastewater treatment systems generally were very effective in reducing the levels of pathogens. USEPA (1999) summarized the effect of biosolids treatment on pathogens (Table 7.14). Further treatment of the solids results in disinfection and destruction of pathogens. Biosolid treatments that achieve a Class A level as defined by USEPA 40 CFR 503 are very effective in eliminating pathogens. USEPA has determined that for Class B biosolids that could harbor pathogens, additional site restrictions are needed to prevent contamination of food, feed and water resources.

The potential for entry of pathogens into the food chain is the ultimate concern of humans and animals. This depends on the survival of pathogens in soil and plants. Furthermore, crop management practices play an important role in preventing contamination and ensuring quality of food and feed products. The next chapter deals with the survival of pathogens in soils and on crops and discusses contamination of air and water resources.

Table 7.14 Effects of Biosolids Treatment on Pathogens[a]

PFRP Treatment	Bacteria	Viruses	Parasites (Protozoa and Helminths)
Anaerobic digestion	0.5–4.0	0.5–2.0	0.5
Aerobic digestion	0.5–4.0	0.5–2.0	0.5
Composting (PFRP)	2.0–4.0	2.0–4.0	2.0–4.0
Air drying	0.5–4.0	0.5–4.0	0.5–4.0
Lime stabilization	0.5–4.0	4.0	0.5

[a]Log reductions shown: A 1-log reduction (tenfold) is equal to a 90% reduction. Class B processes are based on a 2-log reduction.

REFERENCES

Akin, E.W., 1983, Infective dose of waterborne pathogens, Proc. 2nd National Symposium Municipal Wastewater Disinfection, Orlando, Florida, Health Effects Research Laboratory, Cincinnati, OH.

Anonymous, 1975, New chlorine application improves viral kill, *Water Sewage Works*, 30: R-68.

Berg, G. and D. Berman, 1980, Destruction by anaerobic mesophilic and thermophilic digestion of viruses and indicator bacteria indigenous to domestic sludges, *Appl. Environ. Microb.* 39: 361–368.

Besser, R.E., S.M. Lett, J.T. Weber, M.P. Doyle, T.J. Barrett, J.G. Wells and P.M. Griffin, 1993, An outbreak of diarrhea and hemolytic uremic syndrome from *Escherichia coli* 0157:H7 in fresh-pressed apple cider, *JAMA* 269: 2217.

Bovay Engineers, Inc., 1975, Feasibility of land application for Spokane, Washington wastewater solids, Spokane, WA.

Brandon, J.R. and K.S. Neuhauser, 1978, Moisture effects on inactivation and growth of bacterial and fungi in sludges, Sandia Laboratories, Publ. SAND 78–1304, Albuquerque.

Brandon, J.R., W.D. Burge and N.E. Enkiri, 1977, Inactivation by ionizing radiation of *Salmonella enteritidis* serotype Montevideo growth in composted sewage sludge, *Appl. Environ. Microbiol.* 33: 1011–1012.

Bryan, F.L., 1977, Disease transmitted by foods contaminated by wastewater, *J. Food Protection* 40: 45–52.

Burge, W.D. and W.N. Cramer, 1974, Destruction of pathogens by composting sewage sludge, USDA Agricultural Research Service and Maryland Environmental Service and Water Resources Management, Beltsville, MD.

Burge, W.D. and P.D. Millner, 1980, Health aspects of composting: Primary and secondary pathogens, pp. 245–266, G. Bitton, B.L. Damron, G.T. Edds and J.M. Davidson (Eds.), *Sludge — Health Risks of Land Application*, Ann Arbor Science, Ann Arbor, MI.

Burge, W.D., W.N. Cramer and E. Epstein, 1978, Destruction of pathogens in sewage sludge by composting, *Trans. ASAE* 21: 510–514.

Burnham, J.C., 1986, The effect of cement kiln dust and lime on microbial survival in Toledo municipal wastewater sludges, Report from the Department of Microbiology, Medical College of Ohio, Toledo.

Casmore, D.P., 1991, The epidemiology of human cryptosporidiosis and the water route of infection, *Water Sci. Techol.* 24: 157–164.

Casson, L.W., M.O.D. Ritter, L.M. Cossentino and P. Gupta, 1997, Survival and recovery of seeded HIV in water and wastewater, *Water Environ. Res.* 69: 174–179.

Casson, L.W., C.A. Sorber, R.H. Palmer, A. Enrico and P. Gupta, 1992, HIV survivability in wastewater, *Water Environ Res.* 64; 213.

Cave, D.R., 1997, Epidemiology and transmission of *Helicobacter pylori* infection, *Gastroenterology* 113: S9–S14.

Clarke, N., R.E. Stevenson, S.L. Chang and P.W. Kabler, 1961, Removal of enteric viruses from sewage by activated sludge, *Am. J. Public Health* 51: 1118.

Clarke, N., G. Berg, P.W. Kabler and S.L. Chang, 1964, Human enteric viruses in Water: Source, survival and removability, *Advances in Water Pollution Research–Proc. Int. Conf.*, London, England, Pergamon, London.

Clarke, N. and S. Chang, 1975, Removal of enteroviruses from sewage by bench-scale rotary-tube trickling filters, *Appl. Microbiol.* 30: 233.

Cliver, D.O., 1975, Virus association with wastewater solids, *Environ. Lett.* 10: 215–223.

Counts, C.A. and A.J. Shuckrow, 1974, Lime stabilized sludge — its stability and effect on agricultural land, U.S. Environmental Protection Agency, Nat. Environ. Res. Center, Cincinnati, OH.

Cram, E.B., 1943, The effect of various treatment processes on the survival of helminth ova and protozoan cysts in sewage, *Sewage Works,* 15: 1119–1138.

Dahab, M.F., R. Surampalli and P. Ponugoti, 1996, Pathogen and pathogen indicator reduction characteristics in municipal biosolids treatment systems, pp. 265–276, WEFTEC '96, 69th Annual Conference and Exposition, Part I: Residuals and Biosolids Management, Dallas.

Dean, R.S. and J.E. Smith, 1973, The properties of sludge, pp. 39–47, *Proc. Joint Conference on Recycling Municipal Sludges and Effluents on Land,* National Association of State Land-Grant Colleges, Washington, D.C.

Dupont, H.L., C.L. Chappell, C.R. Sterling, P.C. Okhuysen, J.B. Rose and W. Jakubowski, 1995, The infectivity of *Cryptosporidium parvum* in healthy volunteers, *N. Engl. J. Med.* 332: 855–859.

Engineering Science, 1987, Monterey wastewater reclamation study for agriculture, Final Report, Engineering Science, April 1987, Berkeley, CA.

England, B., R.E. Leach, B. Adams and R. Shiosak, 1967, Virologic assessment of sewage treatment at Santee, California, p. 401, in G. Berg (Ed.), *Transmission of Viruses by Water Route*, Interscience, New York.

Enriquez, C.E., C.P. Gerba and M. Abbaszadegan, 1993, Survival of human immunodeficiency virus (HIV) in water and wastewater, p. 859, *Proc. 1993 Water Quality Technol. Conf. Am. Water Works Assoc.*

Epstein, E., 1997, *The Science of Composting*, Technomic, Lancaster, PA.

Epstein, E. and J.F. Donovan, 1992, Pathogens in composting and their fate, *Proc. of Conf. Pathogens in Sludge: What Does it Mean?*, New Orleans, Water Environment Federation.

Epstein, E., J.F. Parr and W.D. Burge, 1977, Health aspects of land application of sewage sludge and sludge compost, National Conference Hazardous Waste Management, San Francisco, Information Transfer, Inc., Rockville, MD.

Farrah, S.R., G. Bitton, E.M. Hoffman, O. Lanni, O.C. Pancorbo, M.C. Lutrick and J.E. Bertrand, 1981, Survival of entericviruses and coliform bacteria in a sludge lagoon, *Appl. Environ. Microbiol.* 41: 459–465.

Farrell, J.B., 1992, Technical support document for reduction of pathogens and vector attraction in sewage sludge, U.S. Environmental Protection Agency, EPA 822/R-93-004, Washington, D.C.

124

LAND APPLICATION OF SEWAGE SLUDGE AND BIOSOLIDS

Farrell, J.B. and G. Stern, 1975, Methods for reducing the infection hazard of wastewater sludge, pp. 19–28, *Radiation for a Clean Environment, Proc. Int. Symp. on the Use of High Level Radiation in Waste Treatment — Status and Prospects*, Munich, Germany.

Farrell, J.B., V. Bhide and J.E. Smith, Jr., 1996, Development of EPA's new methods to quantify vector attraction of wastewater sludges, *Water Environ. Res.* 68: 286–294.

Feachem, R.G., D.J. Bradley, H. Garelick and D.D. Mara, 1980, Appropriate technology for water supply and sanitation: Health effects of excreta and silage management — a state of the art review, World Bank, Washington, D.C.

Fitzgerald, P.R. and R.F. Ashley, 1977, Differential survival of Ascaris ova in wastewater sludge, *J. Water Pollut. Control Fed.* 49: 1722–1724.

Foster, D.H. and R.S. Engelbrecht, 1973, Microbial hazards in disposing of wastewater on soil, pp. 247–270, W.E. Sopper and L.T. Kardos (Eds.), *Recycling Treated Municipal Wastewater and Sludge through Forest and Cropland*, Pennsylvania State University Press, University Park.

Fradkin, L., S.M. Goyal, R.J.F. Bruins, C.P. Gerba, P. Scarpino and J.F. Stara, 1989, Municipal wastewater sludge, The potential public health impacts of common pathogens, *J. Environ. Health* 51: 148–152.

Gaby, W.L., 1975, Evaluation of health hazards associated with solid waste/sewage sludge mixtures, USEPA National Environ. Res. Center, Office of Res. and Dev. EPA-670/2-75-023, Cincinnati, OH.

Gerba, C.P., 1983, Pathogens, pp. 147–195, A.L. Page, T.L. Gleason, J.E. Smith, I.K. Iskander and L.E. Sommers (Eds.), *Utilization of Municipal Wastewater and Sludge on Land*, University of California, Riverside.

Gilbert, J.V., J. Ramakrishna, F.W. Sunderman, Jr., A. Wright and A.G. Plaut, 1995, Protein Hpn: Cloning and characteristics of a histidine-rich metal binding polypeptide in *Helicobacter pylori* and *Heliobacter mustelae, Infect. Immun.* 63: 2682–2688.

Golueke, C.G., 1983, Epidemiological aspects of sludge handling and management, Part II. *BioCycle* 24: 52–58.

Grabow, W.O.K, 1968, The virology of waste water treatment, *Water Res.* 2: 675.

Grubel, P., J.S. Hoffman, F.K. Chong, N.A. Burstein, C. Mepani and D.R. Cave, 1997, Vector potential of houseflies (*Musca domestica*) for *Helicobacter pylori, J. Clin. Microbiol.*

Hornick, R.B., S.E. Greisman, T. E. Woodward, H.L. Dupont, A.T. Dawkins and M.J. Snyder, 1970, Typhoid fever: Pathogenesis and immunological control, *New Engl. J. Med.* 283: 686–691.

Hurst, C., S. Farrah, C. Gerba and J. Melnick, 1978, Development of quantitative methods for the detection of enteroviruses in sewage sludges during activation and following land disposal, *Appl. Environ. Microbiol.* 36: 81–89.

Jewell, W.J., R.M. Kabrick and J.A. Spada, 1980, Autoheated aerobic thermophilic digestion with air aeration, Municipal Research Lab., U.S. Environmental Protection Agency, R804636, Cincinnati, OH.

Johnson, R.W., E.R. Blatchley III and D.R. Mason, 1994, HIV and the bloodborne pathogen regulation: Implications for the wastewater industry, *Water Environ. Res.* 66: 684–688.

Jones, F., A.F. Godfree, P. Rhodes and D.C. Watson, 1983, Salmonellae and sewage sludge — Microbiological monitoring, standards and control in disposing sludge to agricultural lands, pp. 95–114, P.M. Wallis and D.L. Lehmann (Eds.), *Biological Health Risks of Sludge Disposal to Land in Cold Regions*, University of Calgary Press, Alberta.

Kebina, V.Y. and G.L. Ploshcheva, 1974, Sanitary helminthological evaluation of the waste water treatment method on a small industrial sewage plant, *Gig. Sanit.* 7: 93.

Keller, J., 1974, Ozone disinfection pilot plant studies at Laconia, New Hampshire, *J. Am. Water Works Assoc.* 66: 734.

Kelly, H.G. (Ed.), 1991, Autothermal thermophilic aerobic digestion of municipal sludges: Conclusions of a 1-year full scale demonstration project, 64th Annual Conf. of the Water Pollution Control Fed. Toronto.

Kelly, S.M., W.W. Sanderson and C. Neidl, 1961, Removal of enteroviruses from sewage by activated sludge, *J. Water Pollut. Control Fed.* 33: 1056.

Keswick, B.H., 1984, Sources of groundwater pollution, pp. 39–64, G. Bitton and G. Gerba (Eds.), *Groundwater Pollution Microbiology*, John Wiley & Sons, New York.

Knoll, K.H., 1961, Public health and refuse disposal, *Compost Sci.* 2: 35–40.

Koenraad, P.M.F.J., W.F. Jacobs-Reitsma, R.R. Breumer and F.M. Rombouts, 1996, *Campylobacter* spp. in a sewage plant and in the waste water of a connected poultry slaughterhouse, *Water Environ. Res.* 188.

Koenraad, P.M.FJ., F.M. Rombouts and S.H.W. Notermans, 1997, Epidemiological aspects of thermophilic *Campylobacter* in water-related environments: A review, *Water Environ. Res.* 69: 52–63.

Kott, Y., E.M. Nupen and W.R. Ross, 1975, The effect of pH on the efficiency of chlorine disinfection and virus enumeration, *Water Res.* 9: 869.

Kowal, N.E., 1983, An overview of public health effects, pp. 329–394, A.L. Page, T.L. Gleason, J.E. Smith, I.K. Iskander and L.E. Sommers (Eds.), *Utilization of Municipal Wastewater and Sludge on Land*, University of California, Riverside.

Kowal, N.E., 1985, Health effects of land application of municipal sludge, U.S. Environmental Protection Agency, Health Effects Res. Lab., Rep. No. EPA 600/1-85-015, Research Triangle Park, NC.

Krogstad, O. and R. Gudding, 1975, The survival of some pathogenic microorganisms during reactor composting, *Acta Agric. Scandinavia* 25: 281–284.

Laconde, K., R. Lofy and R. Stearns, 1978a, Municipal sludge agricultural utilization practices: An environmental assessment, Vol. I. U.S. Environmental Protection Agency, EPA 530-SW-709, Cincinnati, OH.

Laconde, K., R. Lofy and R. Stearns, 1978b, Municipal sludge agricultural utilization practices, Vol. II, U.S. Environmental Protection Agency, EPA/530/SW-156C, Cincinnati, OH.

Levine, M.M., H.L. Dupont and S.B. Formal, 1973, Pathogenesis of *Shigella dysenteriae* (Shiga) dysentery, *J. Infect. Dis.* 127: 261–270.

Lue-Hing, C., S.J. Sedita and K.C. Rao, 1979, Viral and bacterial levels resulting from land application of digested sludge, pp. 445–462, W.E. Sopper and S.N. Kerr (Eds.), *Utilization of Municipal Sewage Effluent and Sludge on Forest and Disturbed Land*, University Press, College Park, PA.

Majumdar, S.B., W.H. Ceckler and O.J. Sproul, 1974, Inactivation of polio virus in water by ozonation, *J. Water Pollut. Control Fed.* 46: 2048.

Malina, J.F., Jr., 1976, Viral pathogen inactivation during treatment of municipal wastewater, pp. 9–23, L.B. Baldwin, J.M. Davidson and J.F. Gerber (Eds.), *Virus Aspects of Applying Municipal Wastes to Land*, Center for Environmental Programs, Institute of Food and Agricultural Sciences, University of Florida, Gainesville.

Malina, J.F., Jr. and A. Melbard, 1974, Inactivation of virus in bench-scale oxygenated waste stabilization ponds, University of Texas, Austin.

Malina, J.F., Jr., K. Ranganathan, B.P. Sagik and B.E. Moore, 1975, Polio inactivation by activated sludge, *J. Water Pollut. Control Fed.* 47: 2178.

Millner, P.D., K.E. Powers, N.K. Enkiri and W.D. Burge, 1987, Microbial mediated growth suppression and death of salmonella in composted sewage sludge, *Microb. Ecol.* 14: 255–265.

Moore, B.E., B.P. Sagic and C.A. Sorber, 1978, Land application of sludges: Minimizing the impact of viruses on water resources, *Proc. Conf. on Risk Assessment and Health Effects of Land Application of Municipal Wastewater and Sludges*, San Antonio, TX.

Morgan, M.T. and F.W. Macdonald, 1969, Tests show MB tuberculosis doesn't survive composting, *J. Environ. Health* 32: 101–108.

Nath, M.W. and J.C. Johnson, 1979, Quantitative enumeration and evaporation-induced inactivation of enteric viruses in wastewater sludge, Virginia Water Resources Research Center, Virginia Polytechnic Institute, Blacksburg.

Nielsen, A.L. and B. Lydholm, 1980, Methods for the isolation of virus and raw and digested wastewater sludge, *Water Res.* 14: 175–178.

Noland, R.F., J.D. Edwards and M. Kipp, 1978, Full scale demonstration of lime stabilization, U.S. Environmental Protection Agency, EPA-600/2-78-171, Cincinnati, OH.

Novak, J.T., M.P. Eichelberger, S.K. Banerji and J. Yaun, 1984, Stabilization of sludge from an oxidation ditch, *J. Water Pollut. Control Fed.* 56: 950–954.

NRC, 1996, Use of reclaimed water and sludge in food crop production, National Research Council, 1996, Washington, D.C.

NYCDEP, 1992, Sludge management plan, generic environmental impact statement, Chapter IV, Public Health, New York City Department of Environmental Protection, New York.

Pahren, H.R., 1987, Microorganisms in municipal solid waste and public health implications, *CRC Crit. Rev. Environ. Control* 17(3): 187–228.

Parsons, H.R., C. Brownlee, D. Wetter, A. Maurer, E. Haughton, L. Kordner and M. Slezak, 1975, Health aspects of sewage effluent irrigation, Pollution Control Branch, British Columbia Water Resources Service, Victoria.

Pavoni, J. and M. Tittlebaum, 1974, Virus inactivation in secondary wastewater treatment plant effluent using ozone, p. 189, J.F. Malina, Jr. and B.P. Sagik (Eds.), *Virus Survival in Water and Wastewater Systems*, Center for Research in Water Resources, University of Texas, Austin.

Pedersen, D.C., 1981, Density levels of pathogenic organisms in municipal wastewater sludge — a literature review, U.S. Environmental Protection Agency, EPA-600/S2-81-170, Cincinnati, OH.

Pedersen, D.C., A.B. Pincince and J.E. Bates, 1981, Reduction of bacteria, viruses and parasites during conventional sludge treatment processes, Water Pollution Control Federation 54th Annual Conference, Detroit, MI.

Pell, N., 1997, Manures and microbes: Public health and animal health problem? *J. Dairy Sci.* 89: 2673–2681.

Ranganathan, K., J.F. Malina, Jr., and B.P. Sagik (Eds.), 1974, Inactivation of enteric virus during biological wastewater treatment, 7th International Conference on Water Pollution Research, Progress in Water Technology, Pergamon, London.

Reimers, R.S., M.D. Little, D.B. Leftwich, D.D. Bowman, E.J.J. Englande and R.F. Wilkenson, 1980, Parasites in southern sludges and disinfection by standard sludge treatment, National Technical Information Service, EPA Pub. 600/2-81-166. NTIS No. PB82 102344, Springfield, VA.

Reimers, R.S., W.S. Bankston, G.L. Goldstein, Y. Yang and S. Liu, 1996, Disinfection of pathogens by biosolids processing, pp. 51–74, *Stabilization and Disinfection — What Are Our Concerns?*, Water Environment Federation, Dallas, TX.

SAC, 1979, Sewage sludge management program, Vol. 6, Microbiological and virus studies, Sacramento Area Consultants, California.

Sack, R.B., 1975, Human diarrheal disease caused by enterotoxigenic *Escherichia coli, Annu. Rev. Microbiol.* 29: 333–353.

Sekla, L., D. Gemmill, J. Manfreda, M. Lysyk, W. Stackiw, C. Kay, C. Hopper, L. VanBuckenhouat and G. Eibisch, 1980, Sewage treatment plant workers and their environment: A health study, U.S. Environmental Protection Agency, EPA-600/9-80-028, Cincinnati, OH.

Sepp, E., 1980, Pathogen survival in sludge stabilization processes, California Dept. of Health Services, Sacramento.

Sherman, V., 1975, Virus removal in trickling filter plants, *Water and Sewage Works,* April 30: R-36.

Slade, J.S., E.B. Pike, R.P. Eglin, J.S. Colbourne and J.B. Kurtz, 1989, The survival of human immunodeficiency virus in water, sewage and sea water, *Water Sci. Technol.* 21: 55.

Smith, J.E. and J.B. Farrell, 1996, Current and future disinfection — federal perspectives, Water Environ. Fed. 69th Annual Conf. and Exposition, Workshop on Stabilization and Disinfection — What Are Our Concerns? October 5, 1996, Chicago, IL.

Soares, A.C., T.M. Straub, I.L. Petter and C.P. Gerba, 1994, Effect of anaerobic digestion on the occurrence of enteroviruses and giardia cysts in sewage sludge, *J. Environ. Sci. Health A, Environ. Sci. Eng.* 29: 1887–1897.

Sorber, C., B. Moore, D. Johnson, H. Harding and R. Thomas, 1984, Microbial aerosols from application of liquid sludge to land, *J. Water Pollut. Control Fed.* 56: 830.

Stadterman, K.L., A.M. Sninsky, J.L. Sykora and W. Jakubowski, 1995, Removal and inactivation of *Cryptosporidium* oocysts by activated sludge treatment and anaerobic digestion, *Water Sci. Technol.* 31: 97–104.

Stedman's Medical Dictionary, 1977, 23rd ed., Williams & Wilkins, Baltimore, MD.

Stern, G. and J.B. Farrell (Eds.), 1977, Sludge disinfection techniques. *Proc. Nat'l Conf. on Composting of Municipal Residues and Sludges,* Information Transfer, Rockville, MD.

Theis, J.H., V. Bolton and D.R. Storm, 1978, Helminth ova in soil and sludge from twelve U.S. urban areas, *J. Water Pollut. Control Fed.* 50: 2485–2493.

USEPA, 1991, Preliminary Risk Assessment for Parasites in Municipal Sewage Sludge Applied to Land, U.S. Environmental Protection Agency, EPA 600/6-91/001, Washington, D.C.

USEPA, 1992, Technical Support Document for Reduction of Pathogens and Vector Attraction in Sewage Sludge, U.S. Environmental Protection Agency, EPA R-93-004, Washington, D.C.

USEPA, 1999, Control of Pathogens and Vector Attraction in Sewage Sludge, U.S. Environmental Protection Agency, Office of Research and Development, National Risk Management Research Laboratory, Center for Environmental Research Information, EPA/625/R-92-013, Cincinnati, OH.

Vik, T.E. and J.R. Kirk, 1996, Operating experience with the nation's first and world's largest auto-thermal aerobic digestion system, 10th Annual Residual and Biosolids Management Conference: 10 Years of Progress and a Look Toward the Future, Water Environment Federation, Denver, CO.

Walke, R., 1975, The preparation, characterization and agricultural use of bark-sewage compost.

Wang, G., T. Zhao and M.P. Doyle, 1996, Fate of enterohemorrhagic *Escherichia coli* O157:H7 in bovine feces, *Appl. Environ. Microbiol.* 62: 2567.

Ward, R.L. and J.R. Brandon, (Ed.), 1977, Effect of heat on pathogenic organisms found in wastewater sludge, pp. 122–134, Nat'l. Conf. on Composting Municipal Residues and Sludges, Information Transfer, Rockville, MD.

Ward, R.L., G.A. McFeters and J.G. Yeager, 1984, Pathogens in sludge: Occurrence, inactivation and potential regrowth, Sandia National Laboratory, Sand 83–0557.TTC-0428, Albuquerque.

Ward, R.L., D.I. Berstein, E.C. Young, J.R. Sherwood, D.R. Knowlton and G.M. Schiff, 1986, Human rotavirus studies in volunteers: Determination of infectious dose and serological response to infection, *J. Infect. Dis.* 154: 871–880.

Wellings, F.M., A.L. Lewis and C.W. Mountain, 1976, Demonstration of solids-associated virus in wastewater and sludge, *Appl. Environ. Microbiol.* 31: 354–358.

Whitmore, T.N. and L.J. Robertson, 1995, The effect of sewage sludge treatment process on oocysts of *Cryptosporidium parvum*, *J. Appl. Microbiol.* 78: 34–38.

Wiley, J.S., 1962, Pathogen survival in composting municipal wastes, *J. Water Pollut. Control Fed.* 34: 80–90.

Yanko, W.A., 1988, Occurrence of pathogens in distribution and marketing municipal sludges, County Sanitation Districts of Los Angeles County, San Jose Creek Water Quality Laboratory, EPA/600/1-87/014, 1980, Cincinnati, OH.

Yanko, W.A., A.S. Walker, J.J. Jackson, L.L. Libao and A.L. Garcia, 1995, Enumerating salmonella in biosolids for compliance with pathogen regulations, *Water Environ. Res.* 67: 364–370.

CHAPTER 8

Pathogens in Soils and on Plants

INTRODUCTION

The early literature dealt with land application of sewage, effluent, or low-solids sewage sludge (Sepp, 1971). This literature raised concerns about the use of raw sewage, which could result in animal or human health infections (Bicknell, 1972; Dunlop et al., 1951; Dunlop and Wang, 1961). Currently, the practice of applying untreated sewage or sludge is prohibited. The USEPA 503 regulations allow land application of biosolids (treated sewage sludge) either as Class A or Class B material (see Chapter 11 on Regulations).

The existence of pathogens in soils and on plants depends on their survival of the wastewater treatment processes and biosolids treatment, method of land application, soil conditions, and environmental conditions. Although many of the biosolids' processes can result in very effective disinfection and the application of these biosolids does not represent a health hazard, public perception may be sufficiently significant so that applying biosolids to certain food chain crops must be avoided. Figure 8.1 shows the potential routes of pathogen transmission to humans and animals. Food chain crops that are cooked or processed would have no potential for infection. These would include crops for oil (soybeans, sunflower, and canola) or canned foods. Food crops that would not come in contact with biosolids, such as fruit trees (citrus, nut, pears, apples, etc.), would also not harbor pathogens and would not be a potential source of infection. Nonfood chain crops such as fiber (cotton) or forest would also not present a hazard to humans or animals.

Although Class B biosolids can contain pathogens, USEPA provided site restrictions to preclude potential impact to human and animal health and environmental consequences. These site restrictions were based on the potential survival of pathogens in the environment.

When low-solid biosolids are applied by spraying on land, the potential exists for pathogens to be aerosolized. Workers, in particular, may be subjected to these pathogens and become infected. Contamination of water resources, especially drinking water, can result from the movement of pathogens through the soil or in runoff into surface waters.

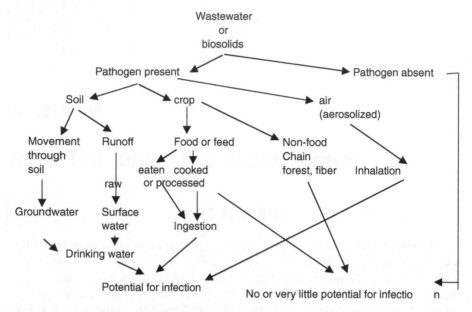

Figure 8.1 Potential routes of pathogen transmission to humans and animals

Since pathogens survive the wastewater treatment processes (primary and second-ary treatment), land application of sewage sludge directly from these processes needs to be avoided or restricted to land management systems. This would ensure minimal potential for environmental contamination or health hazards to humans and animals.

Further treatment, such as digestion, composting, alkaline stabilization, or heat drying, increases the opportunities for land application. Composted products can be used as soil conditioning materials for agricultural and horticultural applications such as landscaping, nurseries, parks, public work projects, and home gardens. Alkaline stabilized biosolids can be used as lime products and soil amendments for agriculture and horticultural practices, such as turf production and certain public works projects. Heat-dried materials are often used as substitutes for chemical fertilizers.

Pathogen survival in soils or plants depends on several factors:

- Climatic and microclimate effects
- Soil physical properties
- Soil chemical properties
- Soil microbial populations
- Plants' exposure to weather

PATHOGENS IN SOILS

The survival and potential movement of pathogens through soils to groundwater depend on several edaphic and climatic factors. The most important are soil moisture, pH, temperature, organic matter, soil texture, soil permeability, sunlight, and antagonistic microflora.

Exposure to sunlight and ultraviolet light will destroy pathogens on the soil surface and on plants. Desiccation and temperature will destroy organisms on the soil surface. Soil physical properties such as soil moisture, soil temperature, organic matter, and ionic strength can affect pathogen survival and movement through soils.

Rudolfs et al. (1950) summarized the literature prior to 1950 on the occurrence and survival of enteric, pathogenic, and relative organisms in soil, water, sewage, sludges, and vegetation. They concluded from the early literature survey that the following soil factors affect the survival of pathogenic microorganisms:

- Type of organism — *E. coli, E. typhosa*, and *M. tuberculosis* appear to be most resistant.
- Soil moisture content — longevity is greater in moist soils.
- High soil moisture holding capacity increases longevity.
- Soil pH — neutral soils favor longevity.
- Organic matter — the type and amount of organic matter may serve as a food and energy source to sustain or allow bacteria to increase.
- The presence of other microorganisms reduces the presence or concentration of pathogenic organisms.
- Temperature — pathogenic organisms survive longer at low temperatures.

Other literature reviews were provided by Dunlop (1968) and Sepp (1971). Gerba et al. (1975) reviewed the literature prior to 1975 and substantiated some findings by Rudolfs et al. (1950) on soil factors. They indicated that sunlight also reduced the survival time on soil surfaces. Reddy et al. (1981) provided a comprehensive review of the behavior and transport of microbial pathogens and indicator organisms in soils treated with organic wastes. They also concluded that the most important factors affecting the die-off rate were temperature, moisture, pH, and method of waste application. Die-off doubled for temperature increases of 10°C, and also rose when soil moisture was reduced. Retention of microorganisms increased with an increase in clay content of the soil. Table 8.1 shows the average die-off rates for selected indicator organisms and pathogens.

The decay die-off rates for *Salmonella typhimurium* and fecal coliform were in the range that Reddy et al. (1981) had shown. The *Shigella sonni* decay coefficient rate was lower. Evens et al. (1995) conducted studies and concluded that *S. sonni* and fecal coliforms were found to survive longer on average than *Salmonella typhimurium*. Casson (1996) also conducted a laboratory study and showed that 2 log reductions were found after 10 to 20 days. Decay rates ranged between 0.08 log per day to 0.4 log per day.

Bacteria

Shortly after the discovery of *Eberthella typhosa* as the causative organism for typhoid, attempts were made to determine its survival in soils. According to Rudolfs et al. (1950), Granger and Deschamps in 1889 claimed to recover the organism from soil after 5.5 years. Also, Karlinski in 1891 reported that *E. typhosa* survived in soil

Table 8.1   Die-off Rate Constants (Day⁻¹) for Selected Indicator Organisms and Pathogens in a Soil-Water-Plant System

Organism	Average	Maximum	Minimum	SD ±	C.V. %	No. of Observations
Escherichia coli	0.92	6.39	0.15	0.64	179	26
Fecal coliforms	1.53	9.10	0.07	4.35	283	46
Fecal streptococci	0.37	3.87	0.05	0.69	188	34
Salmonella sp.	1.33	6.93	0.21	1.70	128	16
Shigella sp.	0.68	0.62	0.74	0.06	9	3
Staphylococcus sp.	0.16	0.17	0.14	0.02	14	2
Viruses	1.45	3.69	0.04	1.44	99	11

Source: Reddy et al., 1981, *J. Environ. Qual.* 10(3): 255–266. With permission.

for 3 months. Melick (1917) reported that *E. typhosa* survived from 29 to 58 days, depending on the soil type. In a sandy soil, survival lasted for 74 days. Kliger (1921) indicated that moist, alkaline conditions in soils were most favorable for the survival of pathogens. Beard et al. (1937) and later Beard himself (1938, 1940) reported that soil water holding capacity, temperature, precipitation, sunlight, soil pH, and soil organic matter all affected the survival of typhoid bacillus. The data showed that the survival of *E. typhosa* was greatest in soils during the rainy season. In sand, where drying is more rapid, the organism survived for a short time — between 4 and 7 days. However, in soils that retained moisture, the organism persisted for longer than 42 days.

Van Dorsal et al. (1967) found a greater die-off rate for *Escherichia coli* and *Streptococcus faecalis* in soil plots exposed to the sun, than those in the shade. The authors also reported that 90% of fecal coliform in the soil were reduced in 3.3 days in the summer and 13.4 days in the winter.

Bacteria can move through soils to great depths. Romero (1970) reported that after 2 days, fecal coliforms and fecal streptococcus organisms were observed to travel over 500 m (1500 ft) after the application of tertiary treated wastewater. This movement occurred in coarse gravel. Several other early authors reported that bacteria could move through soils to depths ranging from <1 m to 830 m. The soils facilitating deep penetration were sand, sandy gravel, and gravel.

The majority of the studies have shown that the movement of bacteria in soils is restricted to less than 30 ft and should not percolate into groundwater (Butler et al., 1954; McGauhey and Krone, 1967). These studies focused primarily on the use of wastewater or very low biosolids concentrations.

Pathogens in dewatered or high-biosolid material applied to land will not likely leach out and move through the surface soil to groundwater. The movement of bacteria where dewatered biosolids are applied is markedly different from with wastewater. Surface application, including tilling or incorporating biosolids into the upper 15 cm (6 in.), greatly reduces the survival and movement of bacteria. Andrews et al. (1983) found that when biosolids were injected into the soil, 90% were inactivated after 17 days in the winter and 3.7 days in the summer.

Sorber and Moore (1986) reviewed the literature prior to 1986 and concluded that quantitative data describing pathogen survival or transport in biosolids-amended soil were extremely limited. Their data are shown in Table 8.2. Generally, some salmonellae bacteria and indicator organisms survived for several weeks. Median die-off rates for indicator bacteria, fecal coliform, fecal streptococci, and total coliform were lower than those observed for *Salmonella*.

The literature review presented by Sorber and Moore (1986) revealed that with one exception, a 90% reduction in *Salmonella* was observed within 3 weeks of biosolids application. These studies were conducted with both indigenous and seeded organisms (Andrews et al., 1983; Jones et al., 1983; Kenner et al., 1971; Larkin et al., 1978). These authors indicate that seeded *Salmonella* organisms often showed higher persistence. They suggested that could be as a result of the very high concentrations, 10^6 to 10^9 per liter, that were land applied. In field studies, indigenous *Salmonella* generally persisted for less than 2 months, with few positive recoveries reported for as long as 3 to 5 months.

Strauch et al. (1981) evaluated the survival of seeded salmonellae in biosolids applied to forest land. The soils were sand and marl. The average temperature was 8.3°C and average precipitation was 739 mm. *Salmonella* survived from 270 to 640 days. Watson (1980), in a study in England, found that organic matter, pH, temperature, and the physical state of the organism affected *Salmonella* survival when digested biosolids was applied to land. *Salmonella* concentrations dropped from 100,000,000 to zero in 42 to 49 days.

The sieving effect of soil, which is influenced by particle size, texture (i.e., clay vs. sand), and adsorption, can greatly reduce bacterial movement. Alexander et al. (1991) studied the factors affecting the movement of bacteria through soil. They measured sorption partition coefficient, hydrophobicity, net surface electrostatic charge, zeta potential, cell size, encapsulation, and flagellation of the cells using 19 different bacterial strains. The results indicated that adsorption greatly contributed to the retention of bacteria, and that bacterial movement through aquifer sand was enhanced by reducing the ionic strength of the in-flowing solution. Cell density and flow velocity also affected bacterial movement. The data revealed that the potential for bacterial contamination of groundwater from the application of biosolids is very minimal. Studies in which bacteria were inoculated in sterile and nonsterile soil showed that pathogens were suppressed by the presence of other soil organisms. Bryanskaya (1964) showed that actinomycetes suppressed the growth of salmonellae and dysentery bacilli.

Pepper et al. (1993) conducted both laboratory and field studies using total coliforms, fecal coliforms, and fecal streptococci organisms. They found that moisture, temperature, and texture of soil affected the survival of these indicator organisms. Survival of organisms was enhanced by increasing soil moisture and clay content and diminished by higher soil temperature. Under field conditions, when rainfall increased soil moisture, regrowth of indicator organisms occurred.

Although concentrations of fecal coliform, fecal streptococci, and *Salmonella* concentrations decreased through an extended hot, dry summer and were not detected, repopulation occurred after precipitation (Gibbs et al., 1997). The authors indicate that despite apparent die-off of salmonellae, land to which biosolids have been applied may be subject to *Salmonella* repopulation. Management needs to take this into account to protect public health.

Table 8.2 Survival of Several Microorganisms in Soil

Organism	Depth (cm)	Die-off Rate - T_{90} Days[1]				Die-off rate T_{99} Days[2]			
		Min.	Max.	Median	Obs.[3]	Min.	Max.	Median	Obs.
Salmonella	0–5	6	61	12	11	11	45	22	8
	5–15	4	22	15	8	7	45	30	6
Fecal streptococci	0–5	7	28	17	10	14	63	24	8
	5–15	NA	NA	NA	NA	NA	NA	NA	NA
Fecal coliform	0–5	7	84	25	19	12	165	60	16
	5–15	4	49	16	10	9	56	32	9
Total coliform	0–5	16	170	40	7	28	350	155	4
	5–15	35	70	42	3	NA	NA	NA	NA
Viruses	0–5	<1	30	3	9	2	52	6	6
	5–15	30	56	30	3	60	100	60	3
Parasites	0–5	17	270	77	11	68	500	81	5
	5–15	NA	NA	NA	NA	NA	NA	NA	NA

[1] T_{90} = 90% reduction within the days indicated.
[2] T_{99} = 99% reduction within the days indicated.
[3] Obs. = Number of observations.
NA = Data not available.
Source: Adapted from Sorber and Moore, 1986.

Land application of biosolids can result in runoff and potential contamination of surface waters. This would be especially true if the biosolids were not incorporated into the soil prior to rainfall (Evens et al., 1995). Land application of biosolids to highly porous soils, following significant amounts of precipitation, could result in some movement of pathogenic organisms for several meters. However, unless the groundwater levels are very shallow, there is little potential for contamination of groundwater. Biosolids application modifies soil properties, which increases the retention and removal of pathogens. Increased organic matter will lower water percolation and enhance water retention in sandy or gravelly soils. Biosolids application modifies pH, which could affect bacterial survival. This is especially true if the pH is increased through liming. The increased organic matter from biosolids' application enhances the indigenous microbial population that could result in pathogen inactivation.

Viruses

Data on viruses in soil from the application of biosolids are meager. However, considerable information is available on viruses from effluent application to land. Viruses in effluents have much greater potential to move through soils and therefore represent much worse scenarios. Movement with liquid media is more rapid than by leaching from a solid matrix. Furthermore, viruses are adsorbed on the solid surfaces and less apt to leach. The organic matter in biosolids would also affect the adsorption of viruses.

The survival of viruses and their movement through soil are greatly affected by soil properties. They generally do not survive very long outside their hosts. They contain a nucleic acid core surrounded by proteins. Viruses are electrically charged colloidal particles and thus capable of adsorbing onto soil surfaces. Many of the studies utilized bacteriophages as models. A bacteriophage is a virus with specific affinity for bacteria (*Stedman's Medical Dictionary*, 1977).

The early studies on the adsorption of viruses onto soil surfaces were reviewed by Bitton (1975). Drewry and Eliassen (1968) studied virus retention in soils and concluded that virus adsorption was affected by the soil-water system pH. Carlson et al. (1968) studied the adsorption of bacteriophage T_2 and type 1 poliovirus to kaolinite, montmorillonite, and illite clays. The type and concentration of cations present in soil water affected sorption of viruses under similar ionic conditions. Kaolinite and montmorillonite adsorbed the same amount of viruses. Illite required twice as much salt to attain the same binding capacity. The authors concluded that the surface exchange capacity, determined by the surface density and clay particle geometry, was important in the adsorption process.

Adsorption is more rapid at a lower pH. Bagdasaryan (1964) reported that enteroviruses survived in loamy and sandy loam soils for prolonged times. The adsorption of viruses to soil may prolong their survival (Gerba et al., 1975). Wellings et al. (1975) indicated that viruses survived in soil for 28 days and were capable of moving through the soil. Tierney et al. (1977) also found that poliovirus type 1 inoculated in raw and activated sludge survived in soils for up to 96 days in the winter. Damgaard-Larsen et al. (1977) discovered that it took 23 weeks during a normal Danish winter to inactivate $10^6 \, TCID_{50}/g$ of Coxsackievirus B3. Under warm, humid conditions in Florida, Farrah et al. (1981) reported a two $\log_{10}$ drop in titer of indigenous viruses in biosolids-amended soil. Gerba (1983) indicated that virus inactivation occurs in the top few centimeters of soil where drying and radiation forces are greatest.

Bitton et al. (1984) evaluated virus transport and survival after land application of biosolids. Strains of poliovirus type 1 and echovirus type 1 were mixed with anaerobically and aerobically digested biosolids and applied to the soil and mixed with the top 2.5 cm of soil. Neither the poliovirus nor echovirus was detectable in soil after being exposed for 8 days to dry fall weather conditions. Under summer weather conditions in Florida, poliovirus was detectable in the soil for 35 days.

Generally, viruses are adsorbed on clays and the adsorption capacity increased with clay content, cation exchange capacity, specific surface, and organic matter. It has been shown that virus adsorption in natural soils follows the Freundlich isotherm:

$q = K_F C^{1/n}$
q = the amount of virus in PFU/g of soil
C = concentration of the virus at equilibrium in PFU/ml solution
K_F = the Freundlich constant determined by the y axis intercept from a plot of q vs. log C
$1/n$ = the slope of the line as determined by the above plot

Burge and Enkiri (1978) showed that the amount of virus adsorbed by five soils was linearly related to the square root of time. There was a high negative correlation with pH, as would be expected, due to their amphoteric nature. The lower the soil pH, the more positively charged the virus particles.

Viruses are removed from percolating water by adsorption on soil particles. Lance et al. (1980) found that poliovirus type 1 essentially remained in the top 5 cm. Other factors are soil type, iron oxides, pH, cations, and virus type. Gilbert et al. (1976) reported that human bacterial and viral pathogens are largely removed as sewage effluent percolates through the soil, and do not move through soil into groundwater. The viruses measured included polio, echo 15, Coxsackie B4, reovirus 1 and 2 and undetermined types. The bacterial indicators and pathogens were: fecal coliforms, fecal streptococci, and *Salmonella* sp. Lue-Hing et al. (1979) did not find any evidence of viral contamination of runoff, surface water, or groundwater at the Fulton County biosolids application site. Although the adsorption of viruses to clays precludes their movement to groundwater, Shaub et al. (1975) have shown that viruses adsorbed are still infectious.

Straub et al. (1992) measured the inactivation rate ($k = \log^{-10}$ reduction per day) of poliovirus type 1 and bacteriophages MS2 and PRD-1 in a laboratory study using two desert soils. Biosolids were added to a Brazito sand loam and Pima clay loam. They found that temperature and soil texture were the most important factors controlling inactivation when the soil was kept moist. As the temperature rose from 15 to 40°C, the inactivation rate for poliovirus and bacteriophage MS2 increased; whereas, for the bacteriophage PRD-1, a significant increase in inactivation occurred only at 40°C. Clay soils afforded more protection than sandy soils to all three viruses. Reduction in moisture content to less than 5% completely inactivated all three viruses within 7 days at 15°C. Thus, a combination of moisture reduction and high temperature is effective in virus inactivation. These studies, under laboratory conditions using soil columns and constant parameters, can point to possible trends, but should not be taken as definitive behavior of organisms in the environment. Soils undergo fluctuations in moisture and temperature. These fluctuations, especially desiccation and high surface temperatures, will destroy pathogens.

Tables 8.3 and 8.4 provide data on the survival of some pathogens in soils.

Table 8.3 Survival of Bacteria in Soils

Organism	Soil	Temperature (°C)	Survival time (days)
Salmonella sp.	Soil	–	15–7,280
Salmonella typhimurium	Soil	Summer sun	<28
	Soil	Summer shade	<70
Salmonella typhi	Soil	–	1–120
	Sandy soil	16–17	<29–<58
	Soil moist	–	<80
	Soil dry	–	<24
Tubercle Bacilli	Soil	–	>189
Leptospira	Soil	–	15–43
Coliform	Soil surface	–	38
Streptococcus spp.	Soil	–	35–63
Fecal streptococci	Soil	–	26–77

Sources: Rudolfs et al., 1950; Parsons et al., 1975; Golueke, 1983.

Parasites

Sorber and Moore (1986) reported that parasites persisted the longest of most organisms in soils. Gerba (1983) found that air, desiccation, and sunlight result in rapid die-off. Protozoan cysts are highly sensitive to drying and are expected to survive in soil for only a few days in most soils. *Entamoeba histolytica* has been reported to survive 18 to 24 h in dry soil and 42 to 72 h in moist soil (Kowal, 1982). Helminth eggs and larvae can survive in soil for longer periods of time. Under favorable conditions of moisture, temperature, and sunlight, *Ascaris, Trichuris*, and *Toxocara* can remain viable and infective for several years (Little, 1980). *Ascaris* eggs can survive up to 7 years in soil (Sepp, 1980). However, Strauch et al. (1981) reported that *Ascaris* eggs are dependent on a host for survival, and thus die fairly quickly in both winter and summer when seeded biosolids are applied to forest land. In an East Bay Municipal Utility District (EBMUD) study, 12.9% of the soil samples contained viable helminth ova 3 years after biosolids' application (Theis et al., 1978). Feachem et al. (1980) reported that hookworms can survive for up to 6 months. Babayera (1966) indicated that *Taenia* could survive from several days to 7 months. Although the data on pathogen survival in soils are highly variable, the evidence suggests that most pathogens do not survive in soils for a great length of time. The soil environment is hostile. Temperature, moisture, and soil pH are the most important factors for pathogen survival. Pathogens die more quickly in warm weather than in colder weather as the soil surface heats up. Decreases in soil moisture hastens pathogen die-off. Soil moisture conditions vary with climate and soil physical properties.

Table 8.4 Survival of Some Viruses in Soil

Organism	Soil	Application System	Temperature (°C)	Survival Time	Reference
Poliovirus	Sand dunes (dry)				
Poliovirus type 1		Effluent		31 days	
			33	40+ days	Palfi, 1972
			30	Reduced by 4 log in 30 days	Moore et al., 1978
Poliovirus 1		Secondary effluent	Winter Summer	89 days 11 days	Tierney et al., 1977
		Activated sewage sludge	Winter Summer	96 days 7 days	Tierney et al., 1977
Coxsackie-virus	Sandy soil and clay	Dewatered, anaerobically digested	Winter	23 weeks	

Table 8.5   Survival of Pathogens or Indicator Organisms on Plants

Organism	Plant	Survival Time	References
Poliovirus 1	Lettuce, radish	23 days	Tierney et al., 1977
Coliforms	vegetables, grass, clover	35 days6–34 days	Parsons et al., 1975
	Tomatoes	35 days	Engelbrecht, 1978
Salmonella typhi	Vegetables	7–53 days	Engelbrecht, 1978
Salmonella typhi	Vegetables and fruit	<1–68 days	Parsons et al., 1975
Shigella spp	Vegetables	2–10 days	Parsons et al., 1975
	Tomatoes	2–7 days	Engelbrecht, 1978
Vibrio cholerae	Vegetables	2 days	Engelbrecht, 1978
Tubercle bacilli	Radish	90 days	Engelbrecht, 1978
	Grass	10–49 days	Parsons et al., 1975
Entamoeba histolytica	Vegetables	3 days	Engelbrecht, 1978
	Vegetables	<1–3 days	Parsons et al., 1975
Tania saginata eggs	Pasture	90–365 days	Engelbrecht, 1978
Enteroviruses	Vegetables and fruit	4–6 days	Parsons et al., 1975

PATHOGENS ON PLANTS

Pathogens do not penetrate into fruits or vegetables unless their skin is broken (Rudolfs et al., 1950; Bryan, 1977). Bryan reviewed the early literature on the survival of pathogens on crops. Pathogen survival would be very short on fruits and vegetables exposed to sunlight and dry conditions. The survival on subsurface crops such as potatoes and beets would be similar to that in soil (Gerba, 1983). The edible portion of crops that does not come in contact with the soil or biosolids is less apt to be contaminated. This is the basis for the Part 503 regulations concerning Class B biosolids used for land application. Table 8.5 provides information on the survival of indicator organisms and pathogens on plants.

Kowal (1982) indicated that survival times of several bacterial pathogens ranged from less than 1 day to 6 weeks. Virus survival on the surface of aerial crops would be expected to be shorter than in soil, because of exposure to deleterious environmental effects, especially sunlight, high temperatures, drying, and washing off by rainfall (Kowal, 1982; Gerba, 1983). Gerba (1983) indicated that the intact surfaces of vegetables are probably impenetrable for viruses. Parasites have been reported to survive on plant surfaces for months.

Sepp (1980) reported that helminth eggs on grass survived over winter and remained infective in harvested hay. He earlier reported (1971) that *Ascaris* ova were destroyed in 27 to 35 days on vegetable surfaces during dry summer weather by desiccation.

In a study by Ohio University and the Ohio Farm Bureau Federation (USEPA, 1985), soil and forage samples were collected on three farms for parasitic ova and larvae. Samples were taken from both biosolids-treated and untreated pasturelands. The analysis was done before biosolids application and 7, 14, and 28 days following application. The study concluded that the risk of parasite transmission to animals from land-applied biosolids was indistinguishable from farms without such application.

Evidence from the literature proves that the survival rate of pathogens on plants is very short. Desiccation, temperature, and ultraviolet light are the most important factors in destroying pathogens on plants. Thus, the risk for humans consuming foods where biosolids are land applied is even lower since most of the biosolids are incorporated into the soil and do not come in contact with edible food crops. The risk to humans from pathogens in biosolids that are applied to non-edible crops, forestry, and fruit trees is essentially nil.

CONCLUSION

The soil environment is generally hostile to pathogens. Desiccation, soil temperature, and pH affect their survival. Decreases in soil moisture have resulted in a greater die-off rate of pathogens. Rises in soil temperature increase the die-off rate of pathogens. Soil acidity also lowers the rate of pathogen survival. Organic matter and clay affect retention, especially of viruses. Generally most pathogens are retained in the upper 5 to 15 cm of the soil.

The data show that bacteria and viruses do not survive in soils for extensive periods. Thus, the potential for bacteria and viruses to move through soils to groundwater is very low. Parasites are larger and heavier than bacteria and viruses and do not readily move through soils. Since most biosolids' applications are immediately incorporated into the soil, the potential for contamination of surface runoff is minimal. The low survivability of pathogens in soils also reduces the potential for surface water contamination.

Today most biosolids are incorporated into the soil before a crop is planted. Exceptions include pastures where liquid biosolids may be applied to an existing crop. Pathogens survive on plants for short periods of time as they are exposed to sunlight and ultraviolet light. Desiccation also plays a major role in reducing their survival.

REFERENCES

Alexander, M., R.J. Wagenet, R.C. Baveye, J.T. Gannon, U. Mingelgrin and Y. Tan, 1991, Movement of bacteria through soil and aquifer sand, U.S. Environmental Protection Agency, Robert S. Kerr Environmental Laboratory, EPA/600/S2-91/010, Ada, OK.

Andrews, D.A., S.L. Mawer and P.J. Matthew, 1983, Survival of salmonellae in sewage sludge injected into soil, *J. Effluent Water Treat.* (G.B.). 23: 72–74.

Babayera, R.I., 1966, Survival of beef tapeworm oncospheres on the surface of the soil in Samarkand, *Med. Parazitiol. Parazit. Bolez.* 35: 557–560.

Bagdasaryan, G.A., 1964, Survival of viruses of the enterovirus group (Poliomyelitis, ECHO and Coxsackie) in soil and on vegetables, *J. Hyg. Epidemiol. Microbiol. Immunol.* 7: 497–505.

Beard, P.J., 1938, The survival of typhoid in nature, *J. Am. Water Works Assoc.* 30: 124.

Beard, P.J., 1940, Longevity of *Eberthella typhosa* in various soils, *Am. J. Public Health.*

Beard, P.J., J.M. Carlson and R.D. Chambers, 1937, The survival of *E. typhosa* in soil, *J. Bacteriol.* 33: 74.

Bicknell, S.R., 1972, *Salmonella aberdeen* in cattle associated with human sewage, *J. Hyg.* 70: 121–126.

Bitton, G., 1975, Adsorption of viruses onto surfaces in soil and water, *Water Res.* 9: 473–484.

Bitton, G., O.C. Pancorbo and S.R. Farrah, 1984, Virus transport and survival after land application of sewage sludge, *J. Appl. Environ. Microbiol.* 47(5): 905–909.

Bryan, F.L., 1977, Disease transmitted by foods contaminated by wastewater, *J. Food Protec.* 40: 45–52.

Bryanskaya, A.M., 1966, Antagonistic effect of actinomycetes on pathogenic bacteria in soil, *Hyg. Sanit.* 31: 123–125.

Burge, W.D. and N.K. Enkiri, 1978, Virus adsorption by five soils, *J. Environ. Qual.* 7: 73–76.

Butler, R.G., G.T. Orlob and P.H. McGauhey, 1954, Underground movement of bacterial and chemical pollutants, *J. Am. Water Works Assoc.* 46: 97–111.

Carlson, J.F.J., F.E. Woodward, D.F. Wentworth and O.J. Sproul, 1968, Virus inactivation on clay particles in natural waters, *Am. J. Public Health* 32: 1256–1262.

Casson, L.W., 1996, Fate and densities of pathogens in biosolids — What are our concerns? pp. 35–36, Water Environment Federation 69th Annual Conference and Exposition, Dallas, TX, October, Water Environment Federation, Alexandria, VA.

Damgaard–Larsen, S., K.O. Jensen, E. Lund and B. Nissen, 1977, Survival and movement of enterovirus in connection with land disposal of sludges, *Water Res.* 11: 503–508.

Drewry, W.A. and R. Eliassen, 1968, Virus movement in groundwater, *J. Water Pollut. Control Fed.* 40: R257–R271.

Dunlop, S.G., 1968, Survival of pathogens and related disease hazards, pp. 107–122, C.W. Wilson and F.E. Beckett (Eds.), *Municipal Sewage Effluent for Irrigation*, Louisiana Tech. Alumni Foundation, Tech. Station, Ruston, LA.

Dunlop, S.G. and W.L. Wang, 1961, Studies on the use of sewage effluent for irrigation of truck crops, *J. Milk Food Technol.* 24: 44–47.

Dunlop, S.G., R.M. Twedt and W.L. Wang, 1951, Salmonella in irrigation water, *Sewage Ind. Waste* 23: 118–122.

Englebrecht, R.S. 1978, Microbial hazards associated with the land application of wastewater and sludge, Ernest Balsom Lecture, University of London, England.

Evens, D.J., L.W. Casson, C.A. Sorber, K.C. Cockley, G. Keleti and B.P. Sagil, 1995, The transport and survivability of selected microorganisms in sludge-amended soils via rainfall-runoff studies, pp. 661–669, Residuals and Biosolids Management *Proc. WEF 68th Annual Conf. and Exposition*, Vol. II, Water Environment Federation, Miami, FL.

Farrah, S.R., P.R. Scheuerman and G. Bitton, 1981, Urea-lysine method for recovery of enteroviruses from sludge, *Appl. Environ. Microbiol.* 41: 459–465.

Feachem, R.G., D.J. Bradley, H. Carelick and D.D. Mara, 1980, Appropriate technology for water supply and sanitation: Health aspects of excreta and silage management — a state of the art review, World Bank, Washington, D.C.

Gerba, C.P., 1983, Pathogens, pp. 147–185, A.L. Page, T.L.I. Gleason, J.E.J. Smith, I.K. Iskandar and L.E. Sommers (Eds.), *Utilization of Municipal Wastewater and Sludge on Land*, University of California, Riverside.

Gerba, C.P., C. Wallis and J.L. Melnick, 1975, Fate of wastewater bacteria and viruses in soil, *ASCE J. Irrigation Drainage* Div. 10: 157–174.

Gibbs, R.A., C.J. Hu, G.E. Ho and I. Unkovich, 1997, Regrowth of faecal coliforms and salmonellae in stored biosolids and soil amended with biosolids, *Water Sci. Technol.* 35(11): 269.

Gilbert, R.G., C.P. Gerba, R.C. Rice, H. Bouwer, C. Wallis and J.L. Melnick, 1976a, Virus and bacteria removal from wastewater by land treatment, *Appl. Environ. Microbiol.* 32: 333–338.

Gilbert, R.G., R.C. Rice, H. Bower, C.P. Gerba, C. Wallis and J.L. Melnick, 1976b, Wastewater renovation and reuse: Virus removal by soil filtration, *Science* 192: 1004–1005.

Golueke, C.G., 1983, Epidemiological aspects of sludge handling and management, Part II. *BioCycle* 24(3): 52–58.

Jones, F., A.F. Godfree, P. Rhodes and D.C. Watson, 1983, Salmonellae and sewage sludge — microbiological monitoring, standards and control in disposing sludge to agricultural lands, 95–114, P.M. Wallis and D.L. Lehmann (Eds.), *Biological Health Risks of Sludge Disposal to Land in Cold Regions*, University of Calgary Press, Alberta.

Kenner, B.A., G.K. Dotson and J.E. Smith, 1971, Simultaneous quantitation of *Salmonella* species and *Pseudomonas aeruginosa:* II. Persistence of pathogens in sludge-treated soils, U.S. Environmental Protection Agency, PB–213 706, National Technical Information Service, Washington, D.C.

Kligler, I.J., 1921, Investigations of soil pollution and the relation of the various privies to the spread of intestinal infections, International Health Board, Rockefeller Institute of Medical Research, Monograph 15, New York.

Kowal, N.E., 1982, Health effects of land treatment: Microbiological, Rep. EPA/600/1-82-007. USEPA Health Effect Res. Lab., Cincinnati, OH.

Lance, J.C. and C.P. Gerba, 1980, Poliovirus movement during high rate land filtration of sewage water, *J. Environ. Qual.* 9: 31–34.

Larkin, E.P., J.T. Tierney, J. Lovett, D. Van Dorsal and D.W. Francis, 1978, Land application of sewage wastes: Potential for contamination of foodstuffs and agricultural soil by viruses, pp. 102–115, B.P. Sagic and C.A. Sorber (Eds.), *Risk Assessment and Health Effects of Land Application of Municipal Wastewater and Sludges*, University of Texas, San Antonio.

Little, M.D., 1980, Agents of health significance: Parasites, pp. 47–58, G. Bitton, B.L. Damron, G.T. Edds and J.M. Davidson (Eds.), *Sludge — Health Risks of Land Application*, Ann Arbor Science, Ann Arbor, MI.

Lue–Hing, C., S.J. Sedita and K.C. Rao, 1979, Viral and bacterial levels resulting from land application of digested sludge, pp. 445–462, W.E. Sopper and S.N. Kerr (Eds.), *Utilization of Municipal Sewage Effluent and Sludge on Forest and Disturbed Land*, University Press, College Park, PA.

McGauhey, P.H. and R.B. Krone, 1967, Soil mantle as a wastewater treatment system, University of California, Sanitary Engineering Research Laboratory Rep. 67–11, Berkeley.

Melick, C.O., 1917, The possibility of typhoid infection through vegetables, *J. Infect. Dis.* 21: 28.

Moore, B.E., B.P. Sagic and C.A. Sorber, 1978, Land application of sludges: Minimizing the impact of viruses on water resources, *Proc. Conf. on Risk Assessment and Health Effects of Land Application of Municipal Wastewater and Sludges,* San Antonio, TX.

Palfi, A., 1972, Survival of enteroviruses during anaerobic digestion, in S.H. Jenkins (Ed.), *Advances in Water Pollution Research, Proc. Sixth Int. Conf.,* Jerusalem, Israel, Pergamon, New York.

Parsons, H.R., C. Brownlee, D. Wetter, A. Maurer, E. Haughton, L. Kordner and M. Slezak, 1975, Health aspects of sewage effluent irrigation, Pollution Control Branch, British Columbia Water Resources Service, Victoria.

Pepper, I.L., K.L. Josephson, R.L. Bailey, M.D. Burr and C.P. Gerba, 1993, Survival of indicator organisms in Sonoran Desert soil amended with sewage sludge, *J. Environ. Sci. Health* A28: 1287–1302.

Reddy, K.R., R. Khaleel and M.R. Overcash, 1981, Behavior and transport of microbial pathogens and indicator organisms in soils treated with organic wastes, *J. Environ. Qual.* 10(3): 255–266.

Romero, J.C., 1970, The movement of bacteria and viruses through porous media, *Groundwater* 8: 37–48.

Rudolfs, W., L.L. Falk and R.A. Ragotzkie, 1950, Literature review of the occurrence and survival of enteric pathogenic and relative organisms in soil, water, sewage and sludge and on vegetation, *Sewage Indust. Wastes* 22: 1261–1281.

Sepp, E., 1971, The use of sewage for irrigation — a literature review, State of California Department Public Health, Bureau of Sanitary Engineering, Sacramento.

Sepp, E., 1980, Pathogen survival in sludge stabilization processes, California Department of Health Services, Sacramento.

Shaub, S.A., E.P. Merer, J.R. Kolmer and C.A. Sorber, 1975, Land application of wastewater: The fate of viruses, bacteria and heavy metals at a rapid infiltration site, Rep. AD–A011263, National Technical Information Service, Washington, D.C.

Sorber, C.A. and B.E. Moore, 1986, Survival and transport of pathogens in sludge–amended soil, 25–32. *Proc. Nat'l. Conf. on Municipal Treatment Plant Sludge Management,* Orlando, FL, Information Transfer, Rockville, MD.

Stedman's Medical Dictionary, 1977, 23rd ed., Williams & Wilkins, Baltimore, MD.

Straub, T.M., I.L. Pepper and C.P. Gerba, 1992, Persistence of viruses in desert soils amended with anaerobically digested sewage sludge, *Appl. Environ. Microbiol.* 58: 636–641.

Strauch, D., W. Konig and F.H. Evers, 1981, Survival of salmonellae and *Ascaris* eggs during sludge utilization in forestry, pp. 408–416, P. L.'Hermité and H. Ott (Eds.), *Characterization Treatment and Use of Sewage Sludge,* Reidel, London.

Theis, J.H., V. Bolton and D.R. Storm, 1978, Helminth ova in soil and sludge from twelve U.S. urban areas, *J. Water Pollut. Control Fed.* 50: 2485–2493.

Tierney, J.T., R. Sullivan and E. Larkin, 1977, Persistence of poliovirus 1 in soil and on vegetables grown in soils previously flooded with inoculated sewage sludge or effluent, *Appl. Environ. Microbiol.* 33: 109–113.

USEPA, 1985, Demonstration of acceptable systems for land disposal of sewage sludge, Water Engineering Research Laboratory, Office of Research and Development, EPA/600/2–86/062, Cincinnati, OH.

Van Dorsal, D.J., E.E. Geldreich and N.A. Clarke, 1967, Seasonal variations in survival of indicator bacteria in soil and their contribution to storm-water pollution, *Appl. Microbiol.* 15: 1362–1370.

Watson, D.C., 1980, The survival of salmonellae in sewage sludge applied to arable land, *J. Water Pollut. Control* 79: 11–18.

Wellings, F.M., A.L. Lewis, C.W. Mountain and L.M. Stark, 1975, Virus consideration in land disposal of sewage effluent and sludge, *Florida Scientist* 38(4): 202–207.

CHAPTER 9

Land Application: Agricultural Crop Responses

INTRODUCTION

The value of sewage sludge and biosolids has been recognized for decades. In 1863, Justice von Liebig in *The Natural Laws of Husbandry* stated:

Even the most ignorant peasant is quite aware that the rain falling upon his dung heap washes away a great many silver dollars and that it would be much more profitable to him to have on his fields what now poisons the air of this house and the streets of the villages; but he looks on unconcerned and leaves matters to take their course, because they have always gone in the same way.

With the advent of sewage treatment in the 1880s, some limited land application occurred. When the activated sludge process was initiated, the use of sludge as organic fertilizer material became of interest (Noer, 1925). Sewage crop irrigation was practiced as early as 1872 in Augusta, Maine (Jewell and Seabrook, 1979).

Extensive research on crop responses to land application of biosolids began in the early 1970s. Much of this research took place in conjunction with an evaluation of heavy metal uptake. This early research included greenhouse studies and field plots. It soon became apparent that greenhouse or pot studies provided very limited information on crop yields, primarily because of the root growth restrictions. However, greenhouse studies did provide trends and directions for field studies. In the greenhouse, the studies could explore many more variations, trials and evaluations. This permitted the researcher to narrow the focus in field experiments.

Considerable research was conducted by the University of Illinois in conjunction with The Metropolitan Sanitary District of Chicago, University of Minnesota, United States Department of Agriculture, Agricultural Research Service, University of California and others. Much research was also conducted by experiment stations under regional auspices, e.g., the regional project W-124 titled: Optimum Utilization of Sewage Sludge on Cropland.

This research was the basis for United States Environmental Protection Agency's (USEPA) early regulations 40 CFR 257, published in 1979 and titled: Criteria for Classification of Solid Waste Disposal Facilities and Practices: Final Interim Final and Proposed Regulations (*Fed. Reg.* 44(179): 53460-53468).

Some of these early publications included:

- *Proceedings Joint Conference Recycling Municipal Sludges and Effluents on Land,* Champaign, Illinois, July 9–13, 1973, National Association of State Universities and Land-Grant Colleges, Washington, D.C.
- *Soils for Management of Organic Wastes and Wastewaters,* 1977, Elliott, L.F. and F.J. Stevenson (Eds.), American Society of Agronomy, Crop Science Society of America and Soil Science Society of America, Madison, WI.
- *Utilizing Municipal Sewage Wastewaters and Sludges on Land for Agricultural Production,* 1977, North Central Regional Extension publication No. 52.
- *Recycling Treated Municipal Wastewater and Sludge through Forest and Cropland,* 1973, Sopper, W.E. and L.T. Kardos (Eds.), Pennsylvania State University Press, University Park.

Page and Chang (1994) provided insight on the growth of technical papers during the period 1971 to 1993. They reported that a casual search of the computer database AGRICOLA showed 876 references from 1970 to March 1993. The greatest number of publications appeared in the 1980s. In 1986 alone, nearly 100 technical papers were published. In 1999, a search of the database Biological Abstracts that cites technical publications worldwide revealed more than 2400 references.

Over the years, numerous changes have occurred that influenced the results observed. Analytical methods have improved, allowing for more accurate results. Improvements in wastewater technology and industrial pretreatment have had probably the greatest influence on biosolids management on crop growth and uptake of trace elements. Major changes have occurred in biosolids' characteristics. Discharges from industrial inputs have been dramatically reduced and domestic plumbing systems shifted from lead and copper to plastic. These factors reduced the input of certain undesirable trace elements. Changes in dewatering systems have also affected the chemical characteristics of biosolids. Vacuum filters using ferric chloride and lime gave way to belt filter presses and centrifuges using polymers.

Several federal laws have also influenced biosolids management. These include:

- The Federal Water Pollution Control Act Amendments of 1972 (Public Law 92-500)
- The Resource Conservation and Recovery Act of 1976 (Public Law 94-580)
- The Marine Protection, Research and Sanctuaries Act of 1972 (Public Law 92-532).

These laws provided grants for the construction of municipal wastewater treatment plants, including sludge processing and management facilities. Public Law 92-500 provided for the pretreatment of toxic industrial wastes prior to discharge into a municipal sewer system, and prohibited the discharge of sewage sludge into navigable waters without a permit from USEPA. Public Law 92-532 restricted and eventually prohibited ocean dumping of sewage sludge. Thus, major metropolitan areas such as New York, Philadelphia, Los Angeles and communities in New Jersey had to seek alternatives such as incineration, land application, composting and heat drying.

The objective of this chapter is to provide data on the effects of land application of biosolids to crops. These effects could be the result of added nutrients and minor plant elements, addition of water under droughty conditions, or improved soil physical properties. Only data on crop productivity are presented. This chapter is divided into sections outlining early research data up until 1970 and later research data, from 1970 to the present.

In the years subsequent to 1970, the quality of biosolids has changed radically. These changes affected plant growth. The dewatering system switched to the use of different chemicals. In the 1970s and early 1980s, ferric chloride and lime were used in dewatering. With the advent of belt filter presses and centrifuges, polymers were used. Industrial pretreatment also resulted in lower heavy metal inputs, lessening the potential for phytotoxicity at high application rates. The application of biosolids containing high levels of phytotoxic heavy metals can result in decreased crop yields (Berti and Jacobs, 1996). Although considerable research has been published outside North America, the data presented here focus on North American studies.

Because much effort was spent on the pros and cons of sludge and biosolids application with respect to heavy metal uptake, many researchers lost sight of the economic value of biosolids application. In many cases if not most, researchers reported only on the chemical constituents of the crops and not on yields. For farmers, the economic benefits of biosolids application are most important, providing the crop is valuable from a health perspective.

AGRONOMIC CROPS

Research Results Prior to 1970

Although for centuries human waste has been applied to soils as a fertilizer for crops, very little has been documented about the benefit of biosolids in agriculture. Bartow and Hatfield (1915) reported that when 1 ton per acre of activated sludge was used, the yield of lettuce increased by 50% and radishes by 300%, compared to check plots. In field trials, Richard and Sawyer (1922) compared the value of activated sludge to that of barnyard manure and ammonium sulfate and reported that the yields of grass, barley and potatoes were similar to the other nitrogen sources. Brown (1921, 1922) also reported that activated sludge was a good source of nitrogen. Noer (1925, 1926) conducted field trials with cabbage, tomatoes, corn and potatoes and found that activated sludge was a satisfactory source of nitrogen.

Lunt (1953) evaluated the potential use of digested biosolids in Connecticut. He found that plant response differed for different soils and that for several plants, liming was needed to obtain a favorable response. He indicated that germination of plants could be inhibited if planting was made directly into soil amended with fresh sewage sludge. Anderson (1959) reported that the use of dried activated sludge as a fertilizer increased from 13,500 to 84,684 metric tons in 1957. During that period, farmers made more modest use of dried digested biosolids and sludge in mixed fertilizers.

In 1967 the University of Illinois initiated a major research effort, sponsored by the Metropolitan Sanitary District of Greater Chicago (Hinesly and Sosewitz, 1969). Initial field studies were conducted on corn and soybean.

Research Results 1970 to 2001

During the early 1970s, the University of Illinois studies applied liquid digested biosolids to several different soils (Lynam et al., 1975). Yields were statistically significantly higher (1% level) than the control in 1970 and 1973, even with lower rates of application. In other years, although there were differences because of the variation in the data, the results were not statistically significant. From 1968 to 1993, the average yield for corn grain in the control was 6.28 metric tons/ha, compared with 8.24 metric tons/ha in field tests with the optimum biosolids application rate.

Milne and Graveland (1972) reported that in a greenhouse using barley, significant increases (1% level) were obtained for barley grown on three soils, even at the rate of 101 metric tons/ha of biosolids. No toxicity was produced even at this high rate. Kelling et al. (1973) applied liquid digested biosolids to the soil surface and obtained increased yields of rye, maize and sorghum Sudan grass. However, they found that spreading biosolids on established alfalfa markedly reduced crown survival and lowered yields. Other glasshouse studies showed that yields of crops were depressed as Cu, Zn and Cr concentrations increased, but Ni had no appreciable effect. It appeared that Cr reduced the effects of the other metals (Cunningham et al., 1975).

Sabey and Hart (1975) also found that severe inhibition of germination of sorghum Sudan grass and millet resulted when seeded shortly after biosolids was incorporated into the soil. They did not find the same inhibition with wheat seeded three months after biosolids application. Giordano et al. (1975) reported that biosolids reduced the yield of mature bush bean pods due to excessive zinc. Forage yields of sweet corn in 1972 and 1973 were higher with biosolids.

One of the most extensive studies, initiated in 1971, was conducted by the USDA-Agricultural Research Service and the University of Minnesota (Linden et al., 1995). When biosolids applications were based on available N and crop-N demands, yields often were higher on biosolid-amended soil than those on areas receiving recommended additions of commercial fertilizer (Dowdy et al., 1976). Dowdy and Larson (1975) reported potato yields of 67.2 metric tons/ha (30 tons/acre) on plots that received 450 metric tons/ha (200 tons/acre) of anaerobically digested biosolids. Yields of 45 metric tons/ha (20 tons/acre) were considered good using commercial fertilizers under the same cultural conditions. They also reported a threefold increase in snap bean yields when biosolids, rather than commercial fertilizers, were used as the source of plant nutrients (see Table 9.1) on coarse sandy loam.

In California, Hyde (1976) reported that anaerobically digested biosolids applied to corn in a field study produced a significantly higher yield than a fertilized control.

Table 9.1   Edible Snap Bean Yields in 1974 as a Function of Biosolids Application

Treatment	Yield - mt/ha
Control	1.2
Fertilized — 675 kg/ha of 8-16-16 & 67 kg/ha N sidedress	5.4
Biosolids application - May 1972	
112 mt/ha	6.2
225 mt/ha	7.9
450 mt/ha	9.3
Biosolids application — May 1972, October 1972 and September 1973 — three equal applications	
337 mt/ha	9.0
675 mt/ha	12.9
1350 mt/ha	16.3

Source: Dowdy and Larson, 1975, *J. Environ. Qual.* 4: 278–282. With permission.

Kelling et al. (1973) applied liquid digested biosolids at rates of 3.75 to 60 metric tons dry solids to rye forage and Sudan-sorghum to two different soils. Yields tended to increase up to 7.5 Mg/ha on the silt loam soil and 15 Mg/ha on the sandy loam soil with dry solids biosolids application. Some crop yield reductions occurred at the 30 and 60 Mg/ha biosolids rates.

Yields of sorghum x Sudan grass hybrid sown directly after biosolids' application increased from 1.93–2.33 tonnes dry matter (DM) applied to soils with no biosolids, to 4.98–571 tonnes with 7.5–30 tonnes of biosolids and declined 4.27–4.33 tonnes with 60 tonnes of biosolids. Clapp et al. (1977) reported that corn yield means for three seasons were 13.8 Mg/ha fodder and 6.4 Mg/ha grain on the fertilizer control area and 14.5 Mg/ha fodder and 6.8 Mg/ha grain on the biosolid-treated area. Reed canarygrass dry matter yields for one cropping season were 7.8 Mg/ha on the fertilized area and 9.7 Mg/ha on the biosolids-treated area.

Considerable research conducted in the 1980s emphasized heavy metal uptake. Watson et al. (1985) reported on cotton yields. As they indicated, both the seed and lint are commercially valuable, but the lint is considered the most valuable. The medium rate of biosolids application (39 Mg/ha) appeared to give the best results during a 3-year study. Essentially, there were no significant differences in seed or lint yield when compared with the fertilized control.

USDA and the University of Minnesota conducted extensive research on the use of biosolids for crop production. Clapp et al. (1986) discovered yield advantages for several crops (see data in Table 9.2). Bidwell and Dowdy (1987), on the other hand, did not find a significant effect on corn stover yields. Corn grain yield from the control was significantly greater than from all the biosolids plots. Heckman et al. (1987) reported that dry matter production of nodulating soybeans was enhanced by application of biosolids, with the greatest increases occurring under moisture stress. However, with biosolids high in heavy metals, it appeared that nodulating dry matter decreased. Crop yield is perhaps the most important determinant of the economic value of biosolids as a substitute for commercial fertilizer (Linden et al., 1995). Corn has value as either a forage crop or a grain crop. Linden et al. (1995) evaluated the effect of biosolids on corn fodder and grain yields over 20 years (see

Table 9.2   Dry Matter Yields of Several Crops from Soil Treated with Biosolids, Compared
with Fertilized Controls in Field Experiments

Crop	Biosolid Application		Yield			References
	Rate (Mg/ha/ Year)	Number of Years	Control (Mg/ha/ Year)	Biosolids (Mg/ha/Year)	Increase (%)	
Corn fodder[1]	116	4	14.9	20.5	38	Clapp et al., 1975
Grain			6.5	9.8	51	
Corn fodder[2]	15	7	16.8	17.4	4	Clapp et al., 1983
Grain			7.9	8.6	9	
Reed canarygrass[2]	18	7	9.8	11.2	14	Clapp et al., 1983
Corn fodder[3]	116	5	17.6	19	8	Clapp et al., 1980
Grain			9.2	9.6	4	
Potato	450	1	1.8[4]	6.7	272	Dowdy et al., 1975
Snap bean	450	3	5.2	16.4	215	Dowdy et al., 1978

[1] Means of 3 years.
[2] Means of 7 years.
[3] Means of 5 years, no additional biosolids.
[4] Unfertilized.
Source: Adapted from Clapp et al., 1986.

Figure 9.1). During that period, the biosolids treatment produced significantly greater fodder yields. Corn grain yields were greater in 15 of the 20 years where biosolids were applied. Over the 20 years, the average corn grain yields were 8.6 Mg/ha (151 bushels per acre) in the biosolids treatments in comparison to 8.1 Mg/ha (140 bushels/acre) in the fertilized control treatments. The average corn fodder yields were 15.8 Mg/ha for the control fertilizer vs. 16.4 Mg/ha for the biosolids treatments.

Berti and Jacobs (1996) evaluated the chemistry and phytotoxicity of soil trace elements as a result of repeated biosolids application. They measured the yield of corn grain, soybean grain and sorghum Sudan grass. The harvest data are shown in Table 9.3. Historically, the use of biosolids has resulted in yields often higher than where recommended applications of fertilizer have been used. In many areas of the United States, farmers continue to use biosolids for their nutrient benefits and increased yields.

FORESTRY AND RECLAMATION

Forestry and reclamation are two excellent uses for biosolids, especially Class B. The major advantages are:

- Use of biosolids in nonfood chain crops
- Revegetation of mined soils
- Reducing runoff and erosion from disturbed soils
- Prevention of surface and groundwater from disturbed soils
- Increased productivity of timber

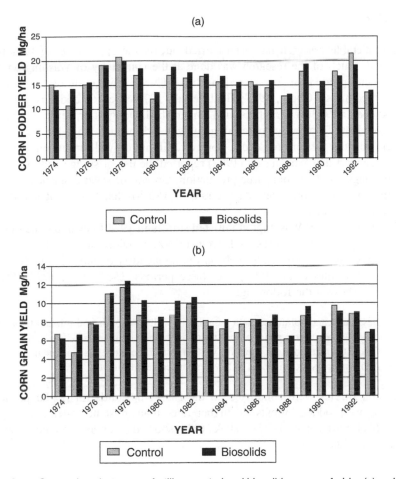

Figure 9.1 Comparison between a fertilizer control and biosolids on corn fodder (a) and grain (b) yield. (After Clapp et al., 1994.)

Table 9.3 Yields of Corn Grain, Soybean Grain and Sorghum-Sudan Grass

| Crop | Control | Biosolids Rate – Mg/ha | | |
		240	870	690
Corn grain[1]	6.1 ± 2.8b[2]	7.6 ± 3.1a	6.1 ± 2.2b	7.7 ± 3.1a
Soybean grain[3]	2.3 ± 1.0a	2.2 ± 1.2a	No plants	2.0 ± 1.1b
Sorghum-Sudan grass[4]	9.1 ± 2.2b	11.0 ± 2.5a	8.4 ± 2.4b	11.4 ± 3.1a

[1] Average of 1985 to 1990 harvests, at 15.5% moisture. Corn was not grown in 1989.
[2] Values in rows followed by the same letter are not significantly different at the P = 0.05 level.
[3] Average of 1985 to 1989 harvests, at 13% moisture.
[4] Average 1985 to 1988 dry weight basis.
Source: Berti and Jacobs, 1996, *J. Environ. Qual.* 25: 1025–1032. With permission.

Forestry

Considerable research has been carried out on the application of biosolids to forestland. Much of the research was done at the University of Washington. Early studies concentrated on forest seedling treated with sewage sludge, biosolids, or biosolids compost (Gouin, 1977; Gouin et al., 1978; Bledsoe and Zasoski, 1981). Zasoski et al., (1983) reported that in a nursery-bed, Douglas fir, Sitka spruce, black cottonwood and Lombardy poplar all grew well in biosolids-amended media. Western hemlock and western red cedar, however, did not do well in biosolids-amended media. Lambert and Weidensaul (1982) reported the use of sludge and biosolids on tree seedlings and Christmas tree production. Growth of several species was better than the control at low rates. At the rate of 180 Mg/ha, transplant survival was significantly reduced.

The University of Washington initiated studies in 1975 to evaluate the effects of biosolids application to forestland. Cole (1982) provided a brief history of the University of Washington's research on the application of biosolids to forestland. One study (Holmes et al., 1993) over three periods, 1985–1987, 1987–1989 and 1989–1991 showed the following:

- Basal area and volume growth increments of the biosolids and thinned treatment outgrew the control and the non-biosolids thinned treatments.
- Quadratic mean diameter growth was largest in the biosolids treatment.
- Applying biosolids appears to have enhanced stand growth.

Several studies were conducted on the use of biosolids for poplar production since poplars produce biomass at higher rates than most north-temperate woody species (Cullington et al., 1997). Zabek (1995) showed that hybrid poplars responded well to biosolids.

Reclamation

One of the best potential uses of biosolids is in reclaiming disturbed lands. Sopper (1993) reported that most of the disturbed lands are from mining of coal, sand, gravel, stone, clay, copper, iron ore, phosphate rock and other minerals. He indicated that 1.6 million ha have been disturbed by surface mining and thousands of acres will be mined annually. Many of these disturbed areas release heavy metals and other contaminants to surface and groundwaters. They are unsightly and are poor habitats to much wildlife.

Using biosolids for mine reclamation offers several major advantages:

- Provides macro- and micronutrients for vegetation establishment
- Improves the soil physical properties, resulting in less erosion and runoff
- Improves the soil biological properties
- Increases land productivity
- Provides food for wildlife

Table 9.4   Reclamation of Disturbed Land

Type of Disturbed Land	State	Biosolids Type	References
Acid strip mine spoil/refuse	PA	Digested dewatered/ composted/liquid	McCormick and Borden, 1973 Kardos et al., 1979 Hill et al., 1979 Murray et al., 1981 Sopper and Kerr, 1982 Sopper et al., 1981 Sopper and Seaker, 1982; 1984 Seaker and Sopper, 1983; 1984; 1987 Dressler et al., 1986 Alberici et al., 1989
Zinc smelter site	PA	Digested dewatered	Sopper, 1989 Sopper and McMahon, 1988
Coal mine spoil	CO	Digested dewatered	Topper and Sabey, 1986 Voos and Sabey, 1987
Copper mine spoil			Sabey et al., 1990
Copper mine	TN	Digested dewatered	Berry, 1982
Strip mine spoils	IL	Digested liquid	Boesch, 1974 Lejcher and Kunkle, 1974 Stucky and Newman, 1977 Hinesley and Redborg, 1984 Stucky et al., 1980 Roth et al., 1982 Joost et al., 1987 Pietz et al., 1989
Strip mine spoil	KY	Digested dewatered	Feuerbacher et al., 1980
	OH	Digested dewatered	Haghiri and Sutton, 1982
	MD	Digested, composted	Griebel et al., 1979
Gravel spoils	MD	Composted	Hornick et al., 1982
C and D canal dredge material	DE	Digested dewatered	Palazzo and Reynolds, 1991
Lignite overburden	TX	Digested dewatered	Cocke and Brown, 1987
Strip mine spoil	VA	Composted	Scanlon et al., 1973
	WV	Digested; digested dewatered, composted	Mathias et al., 1979 Tunison et al., 1982
Iron ore tailings	WI	Digested dewatered	Morrison and Hardell, 1982
Taconite tailings	WI	Digested dewatered	Cavey and Bowles, 1982

Sopper (1993) summarized the research results in land reclamation for the past 20 years. He cited more than 80 publications on the subject. Sopper concluded that the use of municipal sewage biosolids in reclamation and revegetation of drastically disturbed land has been extensively studied. The results to 1993 were encouraging and showed that biosolids, properly applied, can be used to revegetate mined lands in an environmentally safe manner with no major adverse effect on the vegetation, soil, or groundwater. This practice does not pose any significant threat to animal or human health.

This section will not attempt to cover this field since Sopper has done an excellent comprehensive review. Table 9.4 provides several of the citations indicated by Sopper

Table 9.5   Average Dry Matter Production of Several Grasses and Legumes after Three Growing Seasons on Biosolids-Amended Mine Spoil Land

Species	Biosolids Application Rate (Mg/ha)			
	0	40	75	150
	(kg/ha)			
Reed canarygrass	1701	6519	6408	9120
Orchard grass	922	3115	4578	6157
Tall fescue	825	3760	5077	7664
Penngift crownvetch	1937	3283	3326	5070
Bird's-foot trefoil	1586	2712	3731	5518

Sources: Sopper, 1990; 1993.

(1993). Carrello (1990) also provided a 10-year summary on the use of sludge/bio-solids on coal mine spoils.

Sopper (1993) reported that the productivity and fertility of disturbed lands had been substantially improved in most cases by biosolids' application. In West Virginia, where biosolids were applied to mine spoil at rates of 112 and 224 Mg/ha, yield of tall fescue surpassed controls by more than 818% (Mathias et al., 1979).

Sopper (1990) investigated the growth of numerous grasses and legumes on biosolids-amended burned anthracite coal refuse bank. Coverage ranged from 0% to 97%. Excellent coverage was with reed canarygrass, orchard grass and tall fescue, at rates exceeding 40 Mg/ha. Penngift crownvetch averaged more than 68% on plots receiving more than 40 Mg/ha. The average dry matter production for several species is shown in Table 9.5.

CONCLUSION

Scientists have conducted extensive research on the use of biosolids for agricultural crops, horticulture, forestry and reclamation. Biosolids provide macro- and micronutrients. Under certain conditions, the addition of water is also beneficial. The organic matter in biosolids or biosolids products can improve soil physical properties such as soil structure, bulk density, soil moisture, compaction and aeration. These improvements provide for better root growth and development. Plants are better able to utilize plant nutrients and water. In many cases, the use of biosolids increased yield and quality of plants.

Most biosolids are land applied as USEPA Class B materials. These biosolids have restrictions on the type of cropping system where they can be used. Forestry, reclamation and nonfood chain crops (cotton, hay, etc.) are excellent uses for biosolids.

Class A biosolids, such as compost and heat-dried products, are permitted for use on all crops. However, since the additional treatment to produce a Class A is more costly and these products have excellent handling qualities, they are often used in horticulture where crop values are high.

REFERENCES

Alberici, T.M., W.E. Sopper, G.L. Storm and R.H. Yahner, 1989, Trace metals in soil, vege-tation and voles from mine land treated with sewage sludge, *J. Environ. Qual.* 18: 115–120.

Anderson, M.S., 1959, Fertilizing characteristics of sewage sludge, *Sewage Ind. Wastes* 31: 678–682.

Bartow, E. and W.D. Hatfield, 1915, The value of activated sludge as a fertilizer, *Univ. Ill. Bull. Water Survey Ser.* 13: 336–347.

Benchley, W.E. and E.H. Richards, 1920, The fertilizing value of sewage sludges, *J. Soc. Chem. Ind.* 39: 177–182.

Berry, C.R., 1982, Sewage sludge and reclamation of disturbed forest land in the Southeast, pp. 317–320, W.E. Sopper, E.M. Seaker and R.K. Bastian (Eds.), *Land Reclamation and Biomass Production with Municipal Wastewater and Sludge*, Pennsylvania State University Press, University Park.

Berti, W.R. and L.W. Jacobs, 1996, Chemistry and phytotoxicity of soil trace elements from repeated sludge applications, *J. Environ. Qual.* 25: 1025–1032.

Bidwell, A.M. and R.H. Dowdy, 1987, Cadmium and zinc availability to corn following termination of sewage sludge application, *J. Environ. Qual.* 16(4): 438–442.

Bledsoe, C.S. and R.J. Zasoski, 1981, Seedling physiology of eight tree species grown in sludge-amended soil, C.S. Bledsoe (Ed.), *Municipal Sludge Application to Pacific Northwest Forest Lands*, Institute of Forest Resources Contribution No. 41, College of Forest Resources, University of Washington, Seattle.

Boesch, M.J., 1974, Reclaiming the strip mines at Palzo, *Compost Sci.* 15(1): 24–25.

Brown, H.D., 1921, Report on the fertilizer value of activated sludge, *Annu. Rep. Prov. Board of Health of Ontario, Toronto* 40: 117–123.

Brown, H.D., 1922, Report on the fertilizer value of activated sludge, *Annu. Rep. Prov. Board of Health of Ontario, Toronto* 41: 115–126.

Carrello, E.M., 1990, Ten-year summary of environmental monitoring on coal mine spoil amended with sludge, The status of municipal sludge management for the 1990s. *Proc. of the Water Control Federation, 9-1 to 9-19*. Water Control Federation, Alex-andria, VA.

Cavey, J.V. and J.A. Bowles, 1982, Use of sewage sludge to improve taconite tailings as a medium for plant growth, pp. 400–409, W.E. Sopper, E.M. Seaker and R.K. Bastian (Eds.), *Land Reclamation and Biomass Production with Municipal Wastewater and Sludge*, Pennsylvania State University Press, University Park.

Cillington, M., D. Thompson and C. Henry, 1997, The effects of organic residuals on poplars, C.L. Henry and R.B. Harrison (Eds.), *Environmental Effects of Biosolids Management Literature Reviews*, Northwest Biosolids Management Association, Seattle, WA.

Clapp, C.E., R.H. Dowdy, W.E. Larson, D.R. Linden and S.A. Stark, 1975, Liquid sewage sludge as a fertilizer and soil amendment, D1-D17, Utilization of Sewage Wastes on Land, Research progress report, USDA-ARS, University of Minnesota, St. Paul.

Clapp, C.E., D.R. Duncomb, W.E. Larson, D.R. Linden, R.H. Dowdy and R.E. Larson, 1977, Crop yields and water quality after application of sewage sludge to an agricultural watershed, pp. 185–198, R.C. Loehr (Ed.), *Cornell Agricultural Waste Management Conference*, Ithaca, New York, Ann Arbor Science Publishers, Ann Arbor, MI.

Clapp, C.E., W.E. Larson, R.H. Dowdy, D.R. Linden and G.C. Marten, 1983, Utilization of municipal sewage sludge and wastewater effluent on agricultural land in Minnesota, pp. 259–292. *Proc. 2nd Int. Symp. on Peat and Organic Matter in Agric. and Hort.*, Volcani Center, ARO, Bet Dagan, Israel.

Clapp, C.E., S.A. Stark, D.E. Clay and W.E. Larson, 1986, Sewage sludge organic matter and soil properties, Y. Chen and Y. Avnimelech (Eds.), *The Role of Organic Matter in Modern Agriculture*, Martinus Nijhoff, Dordrecht, Netherlands.

Clapp, C.E., Dowdy, R.H., Linden, D.R. et al., 1994, Crop yields, nutrient uptake, soil and water quality during 20 years on the Rosemont sewage sludge watershed, pp. 137–148, C.E. Clapp, W.E. Larson and R.H. Dowdy (Eds.), *Sewage Sludge: Land Utilization and the Environment*, American Society of Agronomy, Crop Science Society of America and Soil Science Society of America, Madison, WI.

Cole, D.W., 1982, Response of forest ecosystems to sludge and wastewater applications — a case study in western Washington, pp. 274–291, W.E. Sopper, E.M. Seaker and R.K. Bastian (Eds.), *Land Reclamation and Biomass Production with Municipal Wastewater and Sludge*, Pennsylvania State University Press, University Park.

Cocke, C.L. and K.W. Brown, 1987, The effect of sewage on the physical properties of lignite overburden. *Reclam. Reveg. Res.* G: 83–93.

Cunningham, J.D., J.A. Ryan and D.R. Keeney, 1975, Phytotoxicity in and metal uptake from soil treated with metal-amended sewage sludge, *J. Environ. Qual.* 4(4): 455–460.

Cunningham, J.D., J.A. Ryan and D.R. Keeney, 1975, Phytotoxicity and metal uptake of metal added to soils as inorganic salts or in sewage sludge, *J. Environ. Qual.* 4(4):460–462.

Dowdy, R.H. and W.E. Larson, 1975, The availability of sludge-borne metals to various vegetable crops, *J. Environ. Qual.* 4: 278–282.

Dowdy, R.H., R.E. Larson and E. Epstein, 1976, Sewage sludge and effluent use in agriculture, in *Land Application of Waste Materials*, Soil Conservation Society of America, Ankeny, IA.

Dressler, R.L., G.L. Strom, W.M. Tzilkowski and W.E. Sopper, 1986, Heavy metals in cottontail rabbits on mined land treated with sewage sludge, *J. Environ. Qual.* 15: 278–281.

Feuerbacher, T.A., R.I. Barnhisel and M.D. Ellis, 1980, Utilization of sewage sludge as a spoil amendment in the reclamation of lands surfaced mined for coal, pp. 187–192, *Proc. Symp. on Surface Mining Hydrol. Sedimentol. and Reclamation*, University of Kentucky, Lexington.

Giordano, P.M., J.J. Mortvedt and A.D. Mays, 1975, Effect of municipal wastes on crop yields and uptake of heavy metals, *J. Environ. Qual.* 4(3): 394–399.

Gouin, F.G., 1977, Conifer tree seedlings respond to nursery soil amended with composted sewage sludge, *Hort. Sci.* 12: 341–342.

Gouin, F.R., C.B. Link and J.F. Kundt, 1978, Forest seedlings thrive on composted sludge, *Compost Sci./Land Util.* 19(4): 28–30.

Griebel, D.A., W.H. Armiger, J.F. Parr, D.W. Steck and J.A. Adam, 1979, Use of composted sewage sludge in revegetation of surface-mined areas, pp. 293–306, W.E. Sopper and S.N. Kerr (Eds.), *Utilization of Municipal Sewage Effluent and Sludge on Forest and Disturbed Land*, Pennsylvania State University Press, University Park.

Haghiri, F. and P. Sutton, 1982, Vegetation establishment on acidic mine spoils as influenced by sludge application, pp. 433–446, W.E. Sopper, E.M. Seaker and R.K. Bastian (Eds.), *Land Reclamation and Biomass Production with Municipal Wastewater and Sludge*, Pennsylvania State University Press, University Park.

Heckman, J.R., J.S. Angle and R.L. Chaney, 1987, Residual effects of sewage sludge on soybeans: I. Accumulation of heavy metals: II. Accumulation of soil and symbiotically fixed nitrogen, *J. Environ. Qual.* 16: 113–124.

Hill, R.D., K. Hinkle and R.S. Klingensmith, 1979, Reclamation of orphaned mined lands with municipal sludges — case studies, pp. 432–443, W.E. Sopper and S.N. Kerr, (Eds.), *Utilization of Municipal Sewage Effluent and Sludge on Forest and Disturbed Land*, Pennsylvania State University Press, University Park.

Hinsley, T.D. and B. Sosewitz, 1969, Digested sludge disposal on crop land, *J. Water Pollut. Control Fed.* 41(5): 822–869.

Hinsley, T.D. and K.E. Redborg, 1984, Long-term use of sewage sludge on agricultural and disturbed lands, Municipal Environmental Research Laboratory, U.S. Environmental Protection Agency, EPA Report 600/2-84-126, Cincinnati, OH.

Holmes, M.J., P. Flemungton-Holmes and D.W. Cole, 1993, Stand growth response to sludge applications at two study sites on the Pack forest, University of Washington, Seattle.

Hornick, S.B., 1982, Crop production on waste amended gravel spoils, pp. 207–218, W.E. Sopper, E.M. Seaker and R.K. Bastian (Eds.), *Land Reclamation and Biomass Production with Municipal Wastewater and Sludge*, Pennsylvania State University Press, University Park.

Hyde, H.C., 1976, Utilization of wastewater sludge for agricultural soil enrichment, *J. Water Pollut. Control Fed.* 48(1): 77–90.

Jewell, W.J., 1979, A history of land application as a treatment alternative, U.S. Environmental Protection Agency, EPA 430/9-79-012, Washington, D.C.

Joost, R.E., F.J. Olsen and J.H. Jones, 1987, Revegetation and minespoil development of coal refuse amended with sewage sludge and limestone, *J. Environ. Qual.* 16: 65–68.

Kardos, L.T., W.E. Sopper, B.R. Edgerton and L.E. DiLissio, 1979, Sewage effluent and liquid digested sludge as aids to revegetation of strip mine spoil and anthracite coal refuse banks, pp. 315–332, W.E. Sopper and S.N. Kerr (Eds.), *Utilization of Municipal Sewage Effluent and Sludge on Forest and Disturbed Land*, Pennsylvania State University Press, University Park.

Kelling, K.A., A.E. Peterson, J.A. Ryan, L.M. Walsh and D.R. Keeney, 1973, Crop responses to liquid digested sewage, pp. 243–252, *Proc. Int. Conf. on Land for Waste Management*, Dept. Environment/National Res. Council, Ottawa, ON.

Lambert, D.H. and C. Weidensaul, 1982, Use of sewage sludge for tree seedling and Christmas tree production, pp. 292–300, W.E. Sopper, E.M. Seaker and R.K. Bastian (Eds.), *Land Reclamation and Biomass Production with Municipal Wastewater and Sludge*, Pennsylvania State University Press, University Park.

Lejcher, T.R. and S.H. Kunkle, 1974, Restoration of acid spoil banks with treated sewage sludge, pp. 184–189, W.E. Sopper and L.T. Kardos (Eds.), *Recycling Treated Municipal Wastewater and Sludge through Forest and Cropland*, Pennsylvania State University Press, University Park.

Linden, D.R., W.E. Larson, R.H. Dowdy and C.E. Clapp, 1995, Agricultural utilization of sewage sludge, Minnesota Agricultural Experiment Station, University of Minnesota, Station Bull. 606–1995, St. Paul, MN.

Lunt, H.A., 1953, The case for sludge as a soil improver, *Water Sewage Works* 100(8): 295–301.

Lynam, B.T., C. Lue-Hing, R.R. Rimkus and F.C. Neil, (Eds.), 1975, The utilization of municipal sludge in agriculture, U.S./Soviet Seminary on *Handling, Treatment & Disposal of Sludges*, Moscow, USSR, The Metropolitan Sanitary District of Greater Chicago.

Mathias, E.L., O.L. Bennett and P.E. Lundberg, 1979, Use of sewage sludge to establish tall fescue on strip mine soils in West Virginia, pp. 307–314, W.E. Sopper and S.N. Kerr (Eds.), *Utilization of Municipal Effluent and Sludge on Forest and Disturbed Land*, Pennsylvania State University Press, University Park.

McCormick, L.H. and F.Y. Borden, 1973, Percolate from spoils treated with sewage effluent and sludge, pp. 239–250, R.J. Hutnick and G. Davis (Eds.), *Ecology and Reclamation of Devastated Land,* Vol. 1, Gordon and Breach, New York.

Milne, R.A. and D.N. Graveland, 1972, Sewage sludge as a fertilizer, *Can. J. Soil Sci.* 52: 270–273.

Morrison, D.G. and J. Hardell, 1982, The response of native herbaceous prairie species on iron-ore tailings under different rates of fertilizer and sludge applications, pp. 410–420, W.E. Sopper, E.M. Seaker and R.K. Bastian (Eds.), *Land Reclamation and Biomass Production with Municipal Wastewater and Sludge,* Pennsylvania State University Press, University Park.

Murray, D.T., S.A. Townsend and W.E. Sopper, 1981, Using sludge to reclaim mine land, *BioCycle* 22(3): 48–51.

Noer, O.J., 1925, Activated sludge: A new source of organic nitrogen, *Am. Fertilizer* 62(12): 24–27.

Noer, O.J., 1926, Activated sludge: Its production, composition and value as a fertilizer, *J. Am. Soc. Agron.* 18(11): 953–962.

Page, A.L. and A.C. Chang, 1994, Overview of the past 25 years: Technical perspective, pp. 3–6, C.E Clapp et al., (Eds.), *Sewage Sludge: Land Utilization and the Environment,* American Society of Agronomy, Crop Science Society of America and Soil Science Society of America, Madison, WI.

Pietz, R.I., J.C.R. Carlson, J.R. Petrerson, D.R. Zenz, and C. Lue-Hing, 1989, Application of sewage sludge and other amendments to coal refuse material: II. Effects on revegetation, *J. Environ. Qual.* 18: 169–173.

Pietz, R.I., J.C.R. Carlson, J.R. Petrerson, D.R. Zenz and C. Lue-Hing, C., 1989, Application of sewage sludge and other amendments to coal refuse materials: I. Effects on chemical composition, *J. Environ. Qual.* 18: 164–169.

Pietz, R.I., J.C.R. Carlson, J.R. Petrerson, D.R. Zenz, and C. Lue-Hing, 1989, Application of sewage sludge and other amendments to coal refuse materials: III. Effects on percolate water composition, *J. Environ. Qual.* 18: 174–179.

Palazzo, A.J. and C.M. Reynolds, 1991, Long-term changes in soil and plant metal concentrations in an acidic dredge disposal site receiving sewage sludge, *Water Air Soil Pollut.* 57–58: 839–848.

Richard, E.H. and G.C. Sawyer, 1922, Further experiments with activated sludge, *J. Soc. Chem. Ind.* 41: 62–72.

Roth, P.L., G.T. Weaver and M. Morin, 1982, Restoration of woody ecosystems on a sludge-amended devastated mine-site, pp. 368–385, W.E. Sopper, E.M. Seaker and R.K. Bastian (Eds.), *Land Reclamation and Biomass Production with Municipal Wastewater and Sludge,* Pennsylvania State University Press, University Park.

Sabey, B.R. and W.E. Hart, 1975, Land application of sewage sludge: I. Effect on growth and chemical composition of plants, *J. Environ. Qual.* 4: 252–256.

Sabey, B.R., R.L. Pendelton and B.L. Webb, 1990, Effect of municipal sewage sludge application on growth of two reclamation shrub species in copper mine spoils, *J. Environ. Qual.* 19: 580–586.

Scanlon, D.H., C. Duggan and S.D. Bean, 1973, Evaluation of municipal compost for strip mine reclamation, *Compost Sci.* 14: 3.

Seaker, E.M. and W.E. Sopper, 1983, Reclamation of deep mine refuse banks with municipal sewage sludge, *Waste Manage. Res.* 1: 309–322.

Seaker, E.M. and W.E. Sopper, 1984, Reclamation of bituminous strip mine spoil banks with municipal sewage sludge, *Reclam. Reveg. Res.* 3: 87–100.

Sopper, W.E.,1989, Revegetation of a contaminated zinc smelter site, *Lands Urban Plan.* 17: 241–250.

Sopper, W.E., 1990, Revegetation of burned anthracite coal refuse banks using municipal sludge, pp. 37–42, *Proc. 1990 National Symposium on Mining*, University of Kentucky, Lexington.

Sopper, W.E., 1993, *Municipal Sludge Use in Land Reclamation*, Lewis Publishers, Boca Raton, FL.

Sopper, W.E. and S.N. Kerr, 1982, Mine land reclamation with municipal sludge — Pennsylvania's demonstration program, pp. 55–74, W.E. Sopper, E.M. Seaker and R.K. Bastian (Eds.), *Land Reclamation and Biomass Production with Municipal Wastewater and Sludge*, Pennsylvania State University Press, University Park.

Sopper, W.E. and J.M. McMahon, 1988a, Revegetation of a superfund site, Part I, *BioCycle* 29(7): 57–60.

Sopper, W.E. and J.M. McMahon,1988b, Revegetation of a superfund site, Part II, *BioCycle* 29(8): 64–66.

Sopper, W.E. and E.M. Seaker, 1982, Strip mine reclamation with municipal sludge, Municipal Environmental Research Lab., U.S. Environmental Protection Agency, Cincinnati, OH.

Sopper, W.E. and E.M. Seaker, 1984, Strip mine reclamation with municipal sludge, Rep. No. EPA-600/2-84-035, U.S. Environmental Protection Agency, Cincinnati, OH.

Sopper, W.E. and E.M. Seaker, 1987, Sludge brings life to microbial communities, *BioCycle* 28(4): 40–47.

Sopper, W.E. and E.M. Seaker, 1987, Development of microbial communities on sludge-amended mine land, pp. 659–680, C.L. Carlson and J.W. Swisher (Eds.), *Innovative Approaches to Mine Land Reclamation*, Southern Illinois University Press, Carbondale.

Sopper, W.E., S.N. Kerr, E.M. Seaker, W.F. Pounds and D.T. Murray (Eds.), 1981, The Pennsylvania program for using municipal sludge for mine land reclamation, pp. 283–290, *Proc. Symp. Surface Mine Hydrol. Sedimentol. and Reclamation*, University of Kentucky, Lexington.

Stucky, D.J. and T.S. Newman, 1977, Effect of dried anaerobically digested sewage sludge on yield and elemental accumulation in tall fescue and alfalfa, *J. Environ. Qual.* 6(3): 271–274.

Stucky, D.J., J.H. Bauer and T.C. Lindsey, 1980, Restoration of acidic mine spoils with sewage sludge, *Reclam. Res.* 3: 129–139.

Topper, K.F. and B.R. Sabey, 1986, Sewage sludge as a coal mine spoil amendment for revegetation in Colorado, *J. Environ. Qual.* 15: 44–49.

Tunison, K.W., B.C. Bearce and H.A. Menser, Jr., 1982, The utilization of sewage sludge: Bark screenings compost for the culture of blueberries on acid minespoil, pp. 195–206, W.E. Sopper, E.M. Seaker and R.K. Bastian (Eds.), *Land Reclamation and Biomass Production with Municipal Wastewater and Sludge*, Pennsylvania State University Press, University Park.

Voos, G. and B.R. Sabey, 1987, Nitrogen mineralization in sewage sludge-amended coal mine spoil and topsoils, *J. Environ. Qual.* 16: 231–237.

Watson, J.E., I.L. Pepper, M. Unger and W.H. Fuller, 1985, Yields and leaf elemental composition of cotton grown on sludge-amended soil, *J. Environ. Qual.* 14(2): 174–177.

Zabek, L.M., 1995, Optimum fertilization of hybrid poplar plantations in coastal British Columbia, Thesis, Dept. of Forestry, University of British Columbia, Vancouver.

Zasoski, R.J., D.W. Cole and C.S. Bledsoe, 1983, Municipal sewage sludge use in forests of the Pacific Northwest, U.S.A.: Growth responses, *Waste Manage. Res.* 1: 103–114.

CHAPTER **10**

Effect of Land Application of Biosolids on Animals and Other Organisms

INTRODUCTION

Land application of biosolids can affect domestic animals and wildlife through the direct ingestion of soil, plant materials, or feed grown on biosolids-amended soil. The U.S Environmental Protection Agency (USEPA), in promulgating the 40 CFR Part 503 regulations, considered the uptake of potentially toxic chemicals or elements that can affect the animal or accumulate in body parts that may become part of the human food chain. Grazing cattle can ingest 1% to 18% of their dry matter intake as soil, and sheep may ingest as much as 30%, depending upon management and the seasonal supply of forage (Fries, 1982; Logan and Chaney, 1983; Harrison et al., 1997). Healy (1968) reported that dairy cattle consumed anywhere from 200 to 600 kg of soil per year, depending on soil and grazing conditions. This could represent approximately 2% of the diet. Chaney (1980) estimated that as much as 6% of the dry matter consumed by cattle in fescue pastures may be biosolids that adhere to the grass, even after several rainfalls. Chaney also indicated that animal grazing studies reveal direct ingestion of biosolids adhering to forage surfaces is the major source of dietary Cd. In promulgating the 40 CFR 503, USEPA evaluated the risk potential to domestic, wildlife, and other organisms from heavy metals as a result of land application of sewage sludge.

The potential impact to animals and other organisms could be the result of the following:

- Pathogens
- Trace elements, especially heavy metals
- Toxic organic compounds

The literature on this subject is rather meager. Land application of biosolids can be beneficial to animals and other organisms or could be potentially harmful, as will be discussed in this chapter.

159

Table 10.1 Mean Apparent Adsorption of Some Heavy Metals by Sheep Fed a Diet Containing Sewage Sludge

Element	Composition in Diet Containing Sewage Sludge (mg/kg)	Control (mg)	Diet Containing Sewage Sludge (mg)
Cd	12.1	ND	+0.6
Cr	143	ND	+10
Cu	912	+1.0	+36
Hg	3.4	+0.073	+0.161
Pb	52	ND	−22.8
Zn	456	−28	−91.7

ND – essentially zero within the limits of detection.

Negative values indicate that more was excreted than adsorbed. No account was provided from domestic water intake.

Source: Smith et al., 1976.

ANIMALS

Domestic

In several studies, domestic animals were fed sewage sludge as part of their daily rations. These represent extreme cases and do not represent normal agricultural practices. However, the results can be useful in understanding the potential impact to animals from consuming large quantities of sludge. USEPA Pathway 10, in the risk assessment of the 40 CFR 503 regulations, offers the closest evaluation of these extreme conditions.

Smith et al. (1976) evaluated the feeding of sewage sludge as a feed supplement to sheep. The apparent absorption of several elements is shown in Table 10.1. The negative adsorption of Pb and Zn indicated that more of the element was excreted than adsorbed. Blood levels did not show appreciable increase in any elements reported. In a subsequent study, cattle were fed sewage sludge solids in supplements. The data on elemental content of kidneys are shown in Table 10.2. There was no difference between the control and the diet containing sewage sludge for Cd, Cr, Hg, Ni, and Zn. Lead was significantly higher from the sludge diet. The authors concluded that the risk to animal health or humans consuming the meat was small.

In the same study, the authors reported on using sewage sludge incorporated into pelleted supplement feed to animals on a range. The results indicated that feeding sewage sludge in a supplement did not cause a significant change in elemental content of blood and livers.

Smith et al. (1978) reported that dietary sewage sludge increased detectable levels of five of 22 pesticides measured in the adipose tissue after 68 days. Levels of Fe and Pb in livers and kidneys were increased as compared to controls, but not beyond levels reported for animals fed conventional diets. In another trial, animals pastured on the range received supplements with cottonseed and sewage sludge. Both supplements increased calf weights over unsupplemented controls.

Table 10.2 Element Content of Kidneys of Cattle as Affected by Sewage Sludge Solids at 20% of Diet after 68 Days

Element	Composition in Diet Containing Sewage Sludge (mg/kg)	Contents of Kidney Control (mg Dry Weight)	Contents of Kidney Diet Containing Sewage Sludge (mg Dry Weight)
Cd	3.5	1.19 ± 0.47	1.49 ± 0.79
Cr	56.5	0.96 ± 0.32	0.65 ± 0.26
Cu	149	18.5 ± 0.80	16.4 ± 0.80
Hg	1.51	0.108 ± 0.075	0.130 ± 0.051
Ni	54	1.60 ± 0.54	1.77 ± 0.53
Pb	81	2.62 ± 0.14	8.43 ± 2.28
Zn	302	90.8 ± 4.5	97.3 ± 10.0

Source: Smith et al., 1978.

One of the earliest comprehensive studies (Baxter et al., 1982) was conducted by the Animal Science Department of Colorado State University in conjunction with the Metropolitan Denver Sewage Disposal District No. 1 (currently Metro Denver Water Reclamation District). This study applied biosolids to an 809-ha (2,000-acre) site from 1969 to 1975. Biosolids filter cake was applied to the surface at the rate of approximately 67 dry Mg/ha (30 dry tons/acre), and plowed under. After a 2-month fallow period, the fields were planted to forage crops, such as winter wheat, oats, sorghum, or Sudan grass.

A herd of 300 to 500 beef cattle continuously grazed the entire area and had ample opportunity to ingest biosolids contaminants. With the exception of Pb, the application of biosolids increased the heavy metal content of the forage. The studies evaluated 12 old cows from the biosolids application site and six cows from a control site where no biosolids were applied. The animals were slaughtered and samples of kidney, liver, bone, muscle, and fat tissues were analyzed for heavy metals. The data are shown in Table 10.3.

The concentration of several refractory organic compounds was also determined. These data, presented in Table 10.4, show no elevated levels of refractory organic compounds in fat tissues.

Biosolids-exposed cattle had higher concentrations of Cd and Zn in kidney and higher Pb concentrations in bone tissue than for the control cattle. These concentrations, however, were within the normal range for cattle of that age. No differences in metal concentrations of muscle tissues existed between the control and biosolids-exposed animals. Friedberg et al. (1974) indicated that Zn may be increased in the organ to counteract toxic effects of Cd.

A significant decrease in Cu content of the liver in the biosolids-exposed cattle was observed, indicating Cu deficiency. The authors found this somewhat unusual; though salt blocks contained low levels of Cu, the forage contained Cu. Mills and Delgarno (1972) indicated that Cd may be an antagonist to Cu.

There were no significant differences in lead concentration in the kidney and liver for the control and exposed animals. The study also evaluated the effect of direct feeding of biosolids as a percentage of the cattle diet.

Table 10.3 Heavy Metal Concentrations in Several Tissues of Biosolids-Exposed Cattle and Control Cattle

Tissue and Group	Heavy Metal (µg/g Dry Weight)								
	As	Cd	Cu	Hg	Mo	Ni	Pb	Se	Zn
Kidney – control	0.03	5.9	16.3	0.12	1.7	<0.5	0.9	5.2	76
Kidney – exposed	0.02	16	16.1	0.06	1.6	<0.5	0.8	5.1	93
Liver – control	0.02	0.8	19	0.02	3.2	<0.8	0.25	1.7	113
Liver – exposed	0.02	1.4	4.6	0.02	2.7	<0.3	0.26	1.0	129
Bone – control	<0.03	0.016	0.5	<0.02	0.15	<0.7	1.5	0.06	58
Bone – exposed	<0.02	0.011	1.3	<0.02	0.35	<0.5	3.1	0.1	68
Muscle – control	<0.02	<0.01	2.9	<0.004	<0.03	<0.2	<0.2	0.67	262
Muscle – exposed	<0.02	<0.04	2.5	<0.004	<0.05	<0.4	<0.2	0.81	247

Source: Baxter et al., 1982, *J. Environ. Qual.* 11: 615–620. With permission.

Table 10.4 Concentration of Refractory Organic Compounds in Fat Tissue of Control and Biosolids-Exposed Cattle

Organic Compound	Control (µg/kg Wet Weight)	Exposed (µg/kg Wet Weight)
Hexachlorobenzene	<10	<10
Alpha-hexachlorocylohexane	10	30
p,p'–DDE	10	10
Dieldrin	10	10
PCBs (Arclor 1254)	500	500

Source: Baxter et al., 1982, *J. Environ. Qual.* 11: 615–620. With permission.

Dowdy et al. (1983) studied the performance of goats and lambs fed corn silage produced on biosolids-amended soil. Cadmium and zinc had increased as a result of biosolids application. Cadmium levels in silage reached a high of 5.26 mg/kg, following an accumulated Cd application of 25.2 kg/ha. Total dry matter intakes by goats did not differ among the control and three biosolids treatments in any of the 3 years. Table 10.5 shows the milk production and feed efficiency for female goats and the feed efficiency for lambs fed corn silage grown on biosolids-amended soils. Dairy milk production for the goats fed silage grown on biosolids amended soils did not differ significantly between breeds, nor between the control and the average of biosolids treatment. Feeding high Cd silage continuously for 3 years did not reduce feed efficiency of dairy goats. Market lambs fed biosolids-fertilized corn silage tended to have higher daily weight gains than the control lambs.

In an earlier study, Heffron et al. 1980) reported that lambs fed corn silage grown on biosolids-amended soil had a lower rate of weight gain than lambs fed control corn silage. The rate of gain was considerably lower than those reported by Dowdy et al. (1983a).

Table 10.5 Performance of Female Goats and Lambs Fed Corn Silage Grown on Biosolids-Amended Soils

Treatment	Goat Milk Reduction kg/day	Goat Feed Efficiency kg milk/kg feed	Lamb Feed Efficiency kg gain/kg intake
Year I			
Control	1.68	1.36	0.15
15 Mg/ha	1.42	1.36	0.18
30 Mg/ha	1.52	1.51	0.21
45 Mg/ha	1.70	1.62	0.19
Year II			
Control	1.53	0.95	0.10
15 Mg/ha	1.07	0.71	0.11
30 Mg/ha	1.42	1.02	0.10
45 Mg/ha	1.31	0.85	0.12
Year III			
Control	1.24	1.04	0.18
15 Mg/ha	0.54	0.20	0.18
30 Mg/ha	1.35	1.04	0.16
45 Mg/ha	1.27	0.91	0.18

Source: Dowdy et al., 1983a, *J. Environ. Qual.* 12(4): 473–478. With permission.

In another study, Dowdy et al. (1983b) examined the milk and blood of goats to evaluate the effect of silage produced on biosolids-amended soil. The data on milk composition of the USEPA-regulated heavy metals are shown in Table 10.6. Other elements, both macro and micronutrients, were also determined. The Cd concentration of goat milk did not increase, even though the animals received as much as 5 mg Cd each day from corn silage containing high levels of bioaccumulated Cd. Zinc concentrations of milk from control animals did not differ from milk from goats fed on silage grown on biosolids-amended soil.

Hill et al. (1998a) assessed the accumulation of potentially toxic elements by direct ingestion of soil and sewage sludge. They fed weaned lambs diets comprising dried grass and various quantities of soil and sludge. There was no limit to the diet. Voluntary intake of dry matter was greatly reduced by the inclusion of sewage sludge in the diet. There was no effect of sewage sludge on digestibility. Live weight gain was depressed. Liver and kidney weights were also reduced. The apparent availability coefficients for Cd, Pb, and Cu rose with increasing levels of sewage sludge. Concentration of these elements increased in the liver and kidney. No increases of Cd and Pb were found in muscle tissue.

In a subsequent paper (Hill et al., 1998b), the authors concluded that accumulation of potentially toxic elements can occur in both the liver and kidney of lambs given grass diets containing elevated levels of Cd derived from sludge-amended soils. The concentrations of Pb in liver and kidney never exceeded the limit for human food of 1 mg/kg fresh weight in any of the treatments. They pointed out that the research also raised some questions. Two soils were from two separate sites. The concentration of Cd in liver and kidney was lower in diets from one site than the

Table 10.6 USEPA Regulated Heavy Metal Content of Milk Collected from Female Goats Fed Corn Silage Grown on Biosolids-Amended Soils

Treatment	Regulated Heavy Metal							
	As	Cd	Cu	Hg	Ni	Pb	Se	Zn
Year I								
Control	<0.03	<0.005	0.42	0.05	<0.15	1.05	<0.13	35.07
30 Mg/ha	<0.03	<0.005	0.32	0.06	<0.16	<1.06	<0.15	37.22
60 Mg/ha	<0.02	<0.004	0.29	0.05	<0.15	1.03	<0.11	33.81
90 Mg/ha	<0.03	<0.003	0.28	0.05	<0.014	<0.85	<0.13	30.76
Year II								
Control	<0.02	0.013	0.79	<0.02	<0.50	<0.50	<0.03	34.07
30 Mg/ha	<0.02	0.011	0.63	<0.02	<0.50	<0.50	<0.06	40.77
60 Mg/ha	—	0.009	0.58	—	<0.50	<0.50	—	34.54
90 Mg/ha	<0.02	0.009	0.53	<0.02	<0.50	<0.50	<0.05	32.85
Year III								
Control	<0.02	0.011	0.64	<0.03	<0.12	<0.33	0.22	39.45
30 Mg/ha	<0.02	0.017	0.43	<0.05	<0.13	<0.35	0.22	40.12
60 Mg/ha	<0.02	0.012	0.26	<0.06	<0.12	<0.31	0.20	34.60
90 Mg/ha	<0.02	0.009	0.29	<0.06	<0.12	<0.33	0.22	36.94

Source: Dowdy et al., 1983b, *J. Environ. Qual.* 12(4): 1983. With permission.

other site at similar levels of total intake of Cd. They concluded that more research is needed on the interaction between chemical speciation of Cd in the soil and availability to the animal.

Again, this is an extreme situation and does not represent what would happen if biosolids were land applied and incorporated into the soil to a depth of 30-cm (6 in.), which is the plow depth. Grazing animals would not consume a diet similar to the feeding experiment. Furthermore, the authors indicated that the sheep did not relish the sludge-treated soil diet. Because biosolids are not applied uniformly in fields, it can be expected that the grazing animals would shy away from heavily biosolid-laden forage.

Very few reports exist on pathogen contamination of domestic animals resulting from sludge or biosolids application. Forbes et al. (1980) reported five outbreaks of cysticerciasis from 1976 to 1979 in cattle from Scotland that were traced to the application of sewage sludge. Some 0.5% of cattle in Scotland are affected. The type applied was liquid undigested sludge.

Wildlife

Numerous studies have been conducted on wildlife coming in contact with soil or crops grown on sludge or biosolids-amended land (Hinsley et al., 1976; Williams et al., 1978; Anderson et al., 1982; Dressler et al., 1986; Hegstrom and West, 1989). Particular concern centers on Cd, because it can accumulate in tissues of herbivorous and omnivorous rodents inhabiting biosolids-amended land. Early studies by West et al. (1981) and Anderson et al. (1982) indicated that the concentrations were not high enough to cause toxicity to wildlife. Beyer (2000) reported that toxicity to wildlife has been exaggerated since threshold data do not support the low levels assumed.

Hinsley et al. (1976) found that, although Zn, Cd, and Ni accumulated in corn as a result of sludge application, only Cd significantly increased in the duodenal, liver, and kidney tissue of ring-necked pheasants fed corn grain from sewage-sludge amended soil. Williams et al. (1978) fed meadow voles crops grown on sludge-amended soil and found significant accumulation of Cd in kidneys and liver, but not muscle. Both of these studies did not simulate normal wildlife conditions. In a field study, in contrast, Alberici et al. (1989) did not find concentrations of Cu, Zn, Co, Cd, and Ni in vole tissues were different from those of the control group. This could be due to differential accumulation of heavy metals by plants, as well as the fact that the animals were not restricted to a specific food source, as was the case for laboratory studies.

Dressler et al. (1986) studied cottontail rabbits on mine land treated with sewage sludge. Liver and muscle tissue of rabbits collected on treated mine sites contained higher levels of Cd than laboratory controls, but overall, cottontail rabbits did not accumulate heavy metals on the sludge-amended site to a level that would be detrimental.

Hegstrom and West (1989) reported the accumulation of heavy metals in small mammals following biosolids application to forest land. Levels of Cd, Pb, Zn, and Cu were measured in the livers and kidneys of insectivorous Trobridge's shrews (*Sorex trobridgii*), shrew-moles (*Neurotrichus gibbsii*), and granivorous deer mice (*Peromyscus maniculatus*) from sludge treated and untreated sites. Heavy metal levels were higher in the livers and kidneys of Trobridge's shrews on the sludge-treated areas than from animals on the untreated sites. Only Cd was elevated in the livers and kidneys of the deer mice. Shrew-moles from the sludge sites had higher levels of Cd and Pb, but not Cu or Zn, compared with animals from the untreated sites.

The effect of Cd concentrations in omnivorous mice (*Peromyscus* sp.) and insectivorous shrews (*Sorex* sp.) inhabiting biosolids-application sites 4, 11 and 15 years after biosolids application was studied by Nickelson and West (1996). They found that kidney Cd was significantly greater in shrews than in mice on all treated sites. Although Cd levels in mice from some treated sites were significantly increased over corresponding controls, levels from the highest application site were comparable to controls found at other sites. They concluded that biosolids would not have a significant long-term effect on omnivores. Cadmium levels in shrews from all treatments, except the low-application rates, were significantly greater than controls. The authors indicated that the levels of Cd did not appear to be biologically significant.

MICROBES

Heavy metals can affect microorganisms (McGrath et al., 1988; Giller et al., 1998). As with many organisms, heavy metals at high concentrations can be toxic. Early evidence shows that near smelters, soil microorganisms and soil microbial populations were impacted by elevated metal concentrations (Freedman and Hutchinson, 1980). Giller et al. (1998) pointed out that early research on biosolids/sewage sludge was focused on protecting against negative effects on crops, on animals grazing on land and on human exposure through the food chain. Some 20 years later, consideration was given to the effect of heavy metals in sewage sludge on microorganisms.

Reddy et al. (1983) studied the survival of *Bradyrhizobium* in sludge-amended soil and found a decline in bacterial populations. They concluded that this decline was the result of heavy metal toxicity. Madariage and Angel (1992) attributed a rapid decline in the population of *Bradyrhizobium* as to toxic concentrations of soluble salts in sludge-amended soil.

Scientists extensively studied the effect of sewage sludge on nitrogen-fixing bacteria. Several early studies were conducted at the Woburn Experiment Station at the Rothamsted Experimental Station in England (McGrath et al., 1988; Giller et al., 1998). They reported a 40% yield reduction in white clover (*Trifolium repens*), compared to field plots receiving farmyard manure, even though the application of organic matter ceased 20 years prior to the study. These authors reported that, in soils receiving sewage sludge over a period of 30 to 50 years, only ineffective *Rhizobium leguminosarium* bv. *trifoli* survived. They concluded that sludge-borne heavy metals, primarily Cd, Zn and Cu, caused these reductions.

Obbard et al. (1993) evaluated the effect of undigested sewage sludge and digested biosolids application on the presence and number of cells of *R. leguminosarium* bv. *trifolii* capable of nodulating white clover. *Rhizobium* were present in all the treatments except low pH soil with the most metal contamination. They concluded that the important factors affecting rhizobial population were soil pH, sludge type, sludge rates, and presence of heavy metals. This data differed from that reported by McGrath (1987). Others (Kinkle et al., 1987) observed an increase in the number of bradyrhizobia in soil, which rose with higher rates of sludge application.

Because of the conflicting data on the effects of heavy metals and the importance of legumes in maintaining soil fertility, Ibekwe et al. (1995) conducted a study to evaluate sewage sludge and heavy metals on nodulation and nitrogen fixation of legumes. Three legumes were studied: alfalfa (*Medicago sativa* L.), white clover (*Trifolium repens* L.) and red clover (*Trifolium pratence* L.). Soil pH and sludge type significantly affected uptake of metals, with phytoxicity observed in low pH sludge-amended soil. Nodulation was reduced but not always completely eliminated in all low pH treatments including controls (soils not receiving sludge). In soils where pH was above 6.0, there was a significant increase in shoot weight and total shoot N with addition of sludge. When pH was maintained at 6.0 or higher, results showed that heavy metals in soil, and resulting increased concentrations in the plant, did not affect nodulation and nitrogen fixation.

The addition of biosolids increases the soil's organic matter and should enhance the soil's microbial population unless sufficient levels of toxins are present that could depress microbial activity. Sastre et al. (1996) found that applications of biosolids at recommended rates increased soil microbial activity. Banerjee et al. (1997) studied the effect of sewage sludge application on biological and biochemical soil properties. Sludge application significantly increased the microbial biomass present in the soils. Sludge application enhanced the N mineralization potential of the soil. Three soil enzymes were monitored and their activities were somewhat enhanced. There was a reduction in microbial diversity. Artiola (1997) reported that denitrification rates in semiarid soils amended with anaerobically digested biosolids were higher within and below the root zone, where biosolids were applied, compared with fertilized soils.

EARTHWORMS

Earthworms are an important component of the natural food chain. They are eaten by birds, snakes, toads, frogs, moles, and centipedes, and these animals in turn are consumed by other animals. The accumulation of heavy metals by earthworms can be toxic to organisms that consume them. Earthworms are also considered a source of protein by some humans. Ireland (1975b) indicated that the metabolic activity of earthworms can increase the solubility of Ca, Pb, and Zn in soils contaminated by heavy metals, and their activity may be important in heavy metal availability to plants.

Earthworms are bioaccumulators of certain heavy metals. Dr. Mike Ireland, a professor of animal physiology at the University College of Wales in Aberystwyth, studied the effect of Pb and Zn in the acid-soil tolerant species *Dendrobaena rubida*. The soil contained over 1000 ppm of Pb and Zn and the concentration of Pb in the earthworm exceeded over 4160 ppm (Ireland, 1975a). In another study conducted later that year, Ireland (1975b) also showed that in soil with 1713 mg/kg Pb the earthworm (*Dendrobaena rubida*) had higher levels of Pb than those in soil with 127 mg/kg Pb.

Helmke et al. (1979) reported that Hg and Cr in the casts of earthworms (*Aporrectodea tuberculata*) rose with increasing biosolids application. However, the constant concentration of these elements in earthworms indicated that these elements are not bioavailable. They also found the earthworms efficiently accumulated Cd. The concentrations of Cd, Cu, and Zn in earthworms rose while the concentration of Se fell with increasing biosolids application. Andersen (1979) also found that Cd increased with sludge application, but there was no clear inverse correlation of sludge application on Cd. In some species, he found that sludge application lowered Cd concentration.

Hartenstein et al. (1980) found that the conversion of waste-activated sludge into egesta by the earthworm *Eisenia foetida* resulted in neither an increase nor decrease of acid-extracted Cd, Ni, Pb, or Zn. The addition of 2500 ppm of copper or copper sulfate to activated sludge killed the earthworms within a week. In a subsequent paper, Hartenstein et al. (1981) concluded that except under extreme conditions, the level of heavy metals in activated sludges will not have an adverse effect on the growth of *E. foetida*.

Beyer et al. (1982) found that earthworms from four sites amended with sludge contained significantly more Cd (12 times), Cu (2.4 times), Zn (2.0 times) and Pb (1.2 times) than did earthworms from control sites. In general, Cd and Zn were concentrated by earthworms relative to soils, and Cu and Pb were not concentrated.

Pietz et al. (1984) determined the concentration of heavy metals in two earthworm species on a biosolids-amended strip mine reclamation site. Although two species, *Lumbricus terrestris* and *Aporrectodea tuberculata* were found on biosolids-amended nonmined fields, only *A. tuberculata* was found in the biosolids-amended and nonamended minesoil fields. Earthworm metal concentrations generally increased over time in all the fields. Metal concentrations in the earthworm accumulated in this order: Cu > Cd > Ni > Cr > Pb > Zn. Earthworm concentrations of Cu and Cd were significantly related to the amounts found in biosolids. This was not true of Ni, Cr and Pb. They concluded that the higher Cd and Cu concentrations in earthworms from biosolids-amended fields might pose a potential hazard to predators.

In Chapter 2 it was shown that the concentration of heavy metals in biosolids has decreased considerably since the 1980s. The lower levels in biosolids today would reduce the concentration of heavy metals in earthworms and lower the potential risk to their predators.

CONCLUSION

The potential toxic effect of land-applied biosolids to both domestic animals and wildlife appears to be very low. Heavy metal concentrations in biosolids during the past 15 years have decreased considerably. Feeding experiments of biosolids-amended soil to animals does not represent the potential intake of heavy metals by grazing animals. The distribution of biosolids on fields is not uniform and today's practices incorporate the biosolids to a depth of 30 cm or more. These practices greatly reduce the potential exposure of grazing animals to potentially toxic elements.

With respect to the effects on the microbial population, the data are very diverse and inconclusive. Some studies showed negative effects and others showed no effects. The studies with earthworms revealed that they are bioaccumulators of heavy metals. Early studies indicated that for some metals the accumulation could represent a risk to predators. As biosolids metal levels decrease, the potential accumulation in the earthworm should decrease, as should the danger to predators.

REFERENCES

Alberici, T.M., W.E. Sopper, G.L. Storm and R.H. Yahner, 1989, Trace metal in soil, vegetation, and voles from mineland treated with sewage sludge, *J. Environ. Qual.* 18: 115–120.

Andersen, C., 1979, Cadmium, lead and calcium content, number and biomass, of earthworms (Lumbricidae) from sewage sludge treated soil, *Pedobiologia* 19: 309–319.

Anderson, T.J., G.W. Barrett, C.S. Clark, V.J. Elia and V.A. Majeti, 1982, Metal concentrations in tissues of meadow voles from sludge-treated fields, *J. Environ. Qual.* 11: 272–277.

Artiola, J.G., 1997, Denitrification activity in the Valdose zone beneath a sludge-amended semi-arid soil, *Commun. Soil Sci. Plant Analysis* 28: 797.

Banerjee, M.R., D.L. Burton and S. Depoe, 1997, Impact of sewage sludge application on soil biological characteristics, *Agric. Ecosystems Environ.* 66: 241–249.

Baxter, J.C., 1983, Heavy metals and persistent organics content in cattle exposed to sewage sludge, *J. Environ. Qual.* 12(3): 316–319.

Baxter, J.C., B. Barry, J.E. Johnson and Kienholz, 1982, Heavy metal retention in cattle tissues from ingestion of sewage sludge, *J. Environ. Qual.* 11: 616–620.

Beyer, W.N., 2000, Hazards to wildlife from soil-borne cadmium reconsidered, *J. Environ. Qual.* 29: 1380–1384.

Beyer, W.N., R.L. Chaney and B.M. Mulhern, 1982, Heavy metal concentration in earthworms from soil amended with sewage sludge, *J. Environ. Qual.* 11(3): 381–385.

Chaney, R.L., 1980, Health risks associated with toxic metals in municipal sludge, pp. 59–83, G. Bitton, B.L. Damron, G.T. Edds, and J.M. Davidson (Eds.), *Sludge — Health Risks of Land Application*, Ann Arbor Science, Ann Arbor, Michigan.

Dowdy, R.H., B.J. Bray, R.D. Goodrich, G.C. Marten, D.E. Pamo and W.E. Larson, 1983a, Performance of goats and lambs fed corn silage produced on sludge-amended soil, *J. Environ. Qual.* 12(4): 473–478.

Dowdy, R.H., B.J. Bray and R.D. Goodrich, 1983b, Trace metal and mineral composition of milk and blood from goats fed silage produced on sludge-amended soil, *J. Environ. Qual.* 12(4): 1983.

Dressler, R.H., G.L. Storm, W.M. Tzilkowski and W.E. Sopper, 1986, Heavy metal in cottontail rabbits on mined land treated with sewage sludge, *J. Environ.Qual.* 15(3): 278–281.

Forbes, G.I., P. Collier, W.J. Reilly and J.C.M. Sharp, 1980, Outbreaks of cysticercosis and salmonellosis in cattle following the application of sewage sludge to grazing, pp. 439–440, *Proc. 1st World Congress on Foodborne Infections and Intoxications,* West Berlin, Germany, Institute of Veterinary Medicine, Berlin.

Freedman, B. and T.C. Hutchinson, 1980, Effect of smelter pollutants on forest litter decomposition near a nickel copper smelter at Sudbury, Ontario, *Can. J. Botany* 58: 1722–1736.

Friedberg, L., M. Piscator, G.F. Nordberg and T. Kjellstrom, 1974, *Cadmium in the Environment,* 2nd ed., CRC, Boca Raton, FL.

Fries, G.F., 1982, Potential polychlorinated biphenyl residues in animal products from application of contaminated sewage sludge to land, *J. Environ. Qual.* 11(1): 14–20.

Giller, K.E., E. Witter and S.P. McGrath, 1998, Toxicity of heavy metals to microorganisms and microbial processes in agricultural soils: A review, *Soil Biol. Biochem.* 30(10/11): 1389–1414.

Harrison, E.Z., M.B. McBride and D.R. Bouldin, 1997, The case for caution recommendations for land application of sewage sludges and an appraisal of the US EPA's Part 503 sludge rules, Cornell Waste Management Institute, Ithaca, NY.

Hartenstein, R., E.F. Neuhauser and J. Collier, 1980, Accumulation of heavy metals in the earthworm *Eisenia foetida, J. Environ. Qual.* 9(1): 23–26.

Hartenstein, R., E.F. Neuhauser and A. Narahara, 1981, Effects of heavy metal and other elemental additives to activated sludge on growth of *Eisenia foetida, J. Environ. Qual.* 10(3): 372–376.

Healy, W.B.,1968, Ingestion of soil by dairy cows, *N.Z. J. Agric. Res.* 11: 487–499.

Heffron, C.L., J.T. Reid et al.,1980, Cadmium and zinc in growing sheep fed silage corn grown on municipal sludge-amended soil, *J. Agric. Food Chem.* 28: 58–61.

Hegstrom, L.J. and S.D. West, 1989, Heavy metal accumulation in small mammals following sewage sludge application to forest, *J. Environ. Qual.* 18: 345–349.

Helmke, P.A., W.P. Robarge, R.L. Korotev and P.J. Schomberg, 1979, Effects of soil-applied sewage sludge on concentrations of elements in earthworms, *J. Environ. Qual.* 8(3): 322–327.

Hill, J., B.A. Stark, J.M. Wilkinson, M.K. Curran, I.J. Lean, J.E. Hall and C.T. Livesey, 1998, Accumulation of potentially toxic elements by sheep given diets containing soil and sewage sludge: I. Effect of type of soil and level of sewage sludge in the diet, *Animal Sci.* 67: 73–86.

Hill, J., B.A. Stark, J.M. Wilkinson, M.K. Curran, I.J. Lean, J.E. Hall and C.T. Livesey, 1998, Accumulation of potentially toxic elements by sheep given diets containing soil and sewage sludge: II. Effect of the ingestion of soils treated historically with sewage sludge, *Animal Sci.* 67: 87–96.

Hinsley, T.D., E.L. Ziegler and J.J. Tyler, 1976, Selected chemical elements in tissues of pheasants fed corn grain from sewage sludge-amended soil, *Agro-Ecosystems* 3: 11–26.

Ibekwe, A.M., J.S. Angle, R.L. Chaney and P. van Berkum, 1995, Sewage sludge and heavy metal effects on nodulation and nitrogen fixation of legumes, *J. Environ. Qual.* 24: 1199–1204.

Ireland, M.P., 1975a, Metal content of *Dendrobaena rubida* (Oligochaea) in a base metal mining area, *Oikos* 26: 74–79.

Ireland, M.P., 1975b, The effect of the earthworm *Dendrobaena rubida* on the solubility of lead, zinc, and calcium in heavy metal contaminated soil in Wales, *J. Soil Sci.* 26: 313–318.

Kinkle, B.K., J.S. Angle and H.H. Keyser, 1987, Long-term effects of metal-rich sewage sludge application on soil populations of *Bradyrhizobium japonicum*, *Appl. Environ. Microbiol.* 53: 315–319.

Logan, T.J. and R.L. Chaney, 1983, Utilization of municipal wastewater and sludges on land — metals. Workshop on Utilization of Municipal Wastewater and Sludge on Land, Denver, CO.

Madariaga, G.M. and J.S. Angle, 1992, Sludge-borne salt effects on survival of *Bradyrhizobium japonicum*, *J. Environ. Qual.* 21: 276–280.

McGrath, S.P., 1987, Long-term studies of metal transfers following application of sewage sludge, pp. 301–317, P.J. Coughtrey, M.H. Martin, and M.H. Unsworth (Eds.), *Pollutant Transport and Fate in Ecosystems*, Blackwell, Oxford, England.

McGrath, S.P., P.C. Brooks and K.E. Giller, 1988, Effects of potentially toxic metals in soils derived from past applications of sewage sludge on nitrogen fixation by *Trifolium repans*, *Soil Biol. Biochem.* 2: 415–424.

Mills, C.J. and A.C. Dalgarno, 1972, Copper and zinc status of ewes and lambs receiving increased dietary concentrations of cadmium, *Nature* 239: 171–173.

Nickelson, S.A. and S.D. West, 1996, Renal cadmium concentrations in mice and shrews collected from lands treated with biosolids, *J. Environ. Qual.* 25: 86–91.

Obbard, J.P., D.R. Sauerbeck and K.C. Jones, 1993, *Rhizobium leguminosarum* bv. *trifoli* in soils amended with heavy metal contamination of sewage sludges, *Soil Biol. Biochem* 25(2): 227–231.

Pietz, R.I., J.R. Peterson, J.E. Prater and D.R. Zenz, 1984, Metal concentrations in earthworms from sewage sludge-amended soils at a strip mine, *J. Environ. Qual.* 13(4): 651–654.

Reddy, G.B., C.N. Cheng and S.J. Dunn, 1983, Survival of *Rhizobium japonicum* in soil, *Soil Biol. Biochem.* 15(3): 343–345.

Sastre, I., M.A. Vicente and M.C. Lobo, 1996, Influence of the application of sewage sludge on soil microbial activity, *Bioresource Tech.* 57: 19–23.

Smith, G.S., H.E. Kiesling, J.M. Cadle, C. Staples and L.B. Bruce, 1976, Recycling sewage solids as feedstocks for livestock, pp. 119–127, *Proc. Third National Conf. Sludge Management Disposal and Utilization,* Miami Beach, FL, Information Transfer, Rockville, MD.

Smith, G.S., H.E. Kiesling and E.E. Ray, 1978, Prospective usage of sewage solids as feed for cattle, pp. 77–102, Sandia Irradiation for Dried Sewage Solids, Seminar Proceedings and Dedication, October 18–19, 1978. Sandia Laboratories, Albuquerque.

West, S.D., R.D. Taber and D.A. Anderson, 1981, Wildlife in sludge-treated plantations, pp. 115–122, C.S. Bledsoe (Ed.), *Municipal Sludge Application to Pacific Northwest Forest Lands,* Bull. 41, College of Forest Resources, University of Washington, Seattle.

Williams, P.H., J.S. Shenk and D.E. Baker, 1978, Cadmium accumulation by meadow voles (*Microtus pennsylvanicus*) from crops grown on sludge-treated soil, *J. Environ. Qual.* 7(3): 450–454.

CHAPTER **11**

Regulations

INTRODUCTION

Biosolids are the only beneficial waste that is regulated by the United States Environmental Protection Agency (USEPA). These regulations pertain to land application of biosolids, including compost and other forms of transformed biosolids materials. States must adhere to the USEPA regulations at a minimum. State agencies may impose more stringent regulations or guidelines. Several agencies in the United States, Canada and Europe have chosen to issue guidelines rather than regulations. Often documents issued as guidelines are used as regulations.

Regulations are important. They provide the public with confidence that the product has met certain criteria and should be safe to use.

The objective of this chapter is to provide current regulations, guidelines and standards prevailing in the United States, Canada and several countries in Europe. This chapter reviews the concepts and approaches leading to regulations and discusses the criteria that should be regulated.

CONCEPTS AND APPROACHES TO REGULATIONS

Kennedy (1992) presented three basic approaches to the development of regulations as related to product use:

- No net degradation
- Risk-based approach
- Best achievable approach

The "no net degradation" concept is based on the premise that the application of biosolids should not increase the level of a heavy metal or other contaminant in the soil. Several European countries and Canadian Provinces have set guidelines or regulations based on this concept. However, no net degradation begs the question: What should be used as a soil base level?

Soil quality varies greatly within a small area; urban soils may have higher levels of lead from leaded gasoline than rural areas. Regional standards would have to be established based on fluctuations in soil quality. If no net degradation were used on a site-by-site basis, it would create excessive sampling requirements and would allow the use of lower quality material on areas that are already contaminated.

Another problem with the no-net-degradation concept: Soils are continuously amended with fertilizers, pesticides, herbicides and other chemicals. This not only changes the baseline quality of the soil, but also illustrates the illogic in singling out a single material as the only regulated material.

The "risk-based approach" considers the potential risk to humans, animals, plants and soil biota, as well as environmental consequences. This approach evaluates the potential toxic effects of a chemical on the individual (human, animal, or plant) or environmental entity. The risk-based approach considers the risk in relation to other risks in the environment. This approach is dependent on having sufficient good data. The most comprehensive risk evaluation focused on heavy metals, resulting in USEPA 40 CFR 503 regulations for the disposal and use of biosolids. This approach was not used for pathogens.

The "best achievable approach" ignores health and environmental aspects and primarily considers technology and economics. Standards are based on what technology can achieve.

United States

U.S. federal regulations dealing with land application of biosolids falls under the jurisdiction of the USEPA. Enforcement is through USEPA regions, with the aid of state regulatory agencies. Those states with delegation have regulatory responsibility.

Regulations promulgated by USEPA cover biosolids or any material containing biosolids. These regulations were required by the Clean Water Act Amendments of 1987 [Sections 405(d) and (e)] as amended (33 U.S.C.A. 1251, *et seq.*). The regulations were published in the *Federal Register* (58 FR 9248 to 9404) as The Standards for the Use or Disposal of Sewage Sludge, Title 40 of the Code of Federal Regulations, Part 503. The 503 rule was published on February 19, 1993 and became effective on March 22, 1993. It was amended on February 25, 1994 (59 FR 9095) for molybdenum. The pollutant concentration limits and annual pollutant loading rates for molybdenum were deleted. Only the ceiling concentration limit of 75 mg/kg was retained.

Two other pollutant limits (for Cr and Se) were contested in the courts. Lawsuits were filed by Leather Industries of America, Inc., Association of Metropolitan Sewerage Agencies, Milwaukee Metropolitan Sewerage District and the city of Pueblo, Colorado. On March 5, 1993, Leather Industries of America filed a petition with the U.S. Circuit Court of Appeals seeking review of the pollutant limits for Cr. Three months later, on June 17, 1993, the City of Pueblo, Colorado filed a petition for review with the U.S. Court of Appeals challenging the Se pollutant limits. On October 25, 1995, USEPA deleted the pollutant limits for Cr and modified the Se limit to 100 mg/kg.

These actions point out important and significant distinctions between regulations and guidelines. Regulations can be overhauled or modified if new data become available or if the regulations are not equally applied. In addition to heavy metals, the 503 rule regulates pathogens and vector attraction. On December 23, 1999 in the *Federal Register* Volume 64, Number 246, pages 72045–72062, USEPA published a proposal to amend the management standards for sewage sludge. A numeric concentration limit is proposed for dioxin and dioxin-like compounds in sewage sludge that is applied to the land, as well as monitoring, record keeping and reporting requirements for dioxins in land-applied sewage sludge.

Much of the discussion in this chapter is from four USEPA documents.

1. *Federal Register.* Friday February 19, 1993. *Standards for the Use or Disposal of Sewage sludge; Final Rules. Part II* Environmental Protection Agency. 40 CFR Part 257.
2. USEPA. Office of Wastewater Management (4204). A Plain English Guide to the EPA Part 503 Biosolids Rule. EPA/832/R-93/003. September 1994.
3. USEPA. Office of Wastewater Management (4204). Guide to the Biosolids Risk Assessments for the Part 503 Rule. EPA832-B-95-005. Unpublished document. Courtesy of Dr. J. Walker.
4. USEPA. Office of Research and Development. Environmental Regulations and Technology. Control of Pathogens and Vector Attraction in Sewage Sludge. EPA/625/R-92/013. Revised October 1999. Washington, D.C.

The 503 rule was designed to protect public health and the environment from "any reasonably anticipated adverse effects of certain pollutants and contaminants that may be present in [biosolids]" (USEPA, 1994). USEPA clearly stated that it promotes the beneficial use of biosolids. A very intensive risk assessment was conducted. The rule-making took 9 years and evaluated research from the previous 25 years. In 1984 USEPA considered 200 pollutants identified in the "40 Cities Study." The selection of the 200 pollutants was based on the following criteria:

- Human exposure and health effects
- Plant uptake of pollutants
- Phytotoxicity
- Effects in domestic animals and wildlife
- Effects in aquatic organisms
- Frequency of pollutant occurrence in biosolids

This list of pollutants was submitted for review by four panels. The panels recommended that approximately 50 of the 200 pollutants listed be further studied. In the final regulations, USEPA addressed 24 pollutants using 14 exposure pathways (Ryan and Chaney, 1995). The 24 pollutants were:

Organics	Heavy Metals
Aldrin/dieldrin (total)	Arsenic
Benzene	Cadmium
Benzo(a)pyrene	Chromium
Bis(2-ethylhexyl)phthalate	Copper
Chlordane	Lead
DDT/DDE/DDD (total)	Mercury
Heptachlor	Molybdenum
Hexachlorobenzene	Nickel
Hexachlorobutadiene	Selenium
Lindane	Zinc
N-Nitrosdimethylamine	
Polychlorinated biphenyls	
Toxaphene	
Trichloroethylene	

Risk assessment followed four basic steps (USEPA, 1995).

- Hazard identification: Can the identified pollutants harm human health or the environment?
- Exposure assessment: Who is exposed, how do they become exposed and how much exposure occurs? Highly exposed individuals were identified and their exposure to pollutants in biosolids evaluated. Fourteen exposure pathways were identified for land
  application of biosolids (see Table 11.1).
- Dose–response evaluation. What is the likelihood of an individual developing a particular disease as the dose and exposure increases? These two EPA toxicity factors were used whenever available:
  — Risk reference doses (RFDs) — daily intake
  — Cancer potency values (q_1*s) — conservative indication of the likelihood of a chemical inducing or causing cancer during the lifetime of a continuously exposed individual.
- Risk characterization: What is the likelihood of an adverse effect in the population exposed to a pollutant under the conditions studied? Risk is calculated as: Risk = Hazard × Exposure. Hazard refers to the toxicity of a substance determined during the hazard's identification and dose–response evaluation and exposure is determine through the exposure assessment (USEPA, 1995). EPA made a policy decision to regulate risk at 1×10^{-4}.

The general approach USEPA utilized in developing pollutant soil loading limits follows (Ryan and Chaney, 1995):

- Delineation of pollutants of concern in biosolids.
- Identification of potential pathways for exposure and receptors (humans, soil biota, plants and animals) to several pollutants through land application of biosolids.
- Identification of dose–response relationships for the receptors and pollutants of concern.

- Determination of maximum acceptable loading rates of biosolids to land for each pollutant based on the most limiting value for all evaluated pathways.
- Determination of the pollutant limits (cumulative soil pollutant application limit and maximum allowed biosolids pollutant concentration). This was obtained from maximum loading rates and biosolids concentration from the National Sewage Sludge Survey.

Several key assumptions were used in determining the pollutant limits:

- The target organism was the highly exposed individual (HEI) rather than the most exposed individual (MEI). The HEI was a realistic individual, whereas the MEI was unrealistic and did not exist.
- EPA used the lifetime exposure criteria of 70 years. For home gardeners producing their own food, it was assumed that 59% of the food would be grown in home gardens amended with biosolids.
- Uptake slopes for pollutants by crops were assumed to be linear even though the data indicated a curvilinear slope. This was believed to be more conservative.
- Cancer risk for all biosolids use was set at 1×10^{-4}.
- Data for plant uptake were based on field data when available.
- Human dietary exposure to pollutants in biosolids was revised from the early assessment by apportioning food consumption among several different age periods during the lifetime of the 70 years of the HEI.
- The final rule evaluated all organic pollutants proposed for the 503 rule. The levels found by the National Sewage Sludge Survey showed that organic pollutants were at low levels and in the evaluation did not pose significant risks to public health or the environment. USEPA is currently considering a zero limit for PCBs.

Examples of the risk assessment and the determination of the pollutant limits are shown for arsenic. The first analysis is for Pathway 1, where, over a lifetime, an adult consumes crops grown on biosolids-amended soil. The second example uses Pathway 3, a child ingesting biosolids. Based on these analyses it was determined that Pathway 3 was the limiting pathway. These analyses are based on USEPA (1995):

$$RIA = \frac{RfD*BW}{RE} TBI*10^3 \qquad (11.1)$$

where

RIA   = allowable dose of pollutant without adverse effects
RfD   = reference dose in mg/kg-day; for As = 0.0008 mg/kg-day
BW   = human body weight, 70 kg
RE   = relative effectiveness of ingestion exposure, 1.0, no units
TBI   = total pollutant intake from all background sources in water, food and air, 0.012 mg/day

For arsenic in biosolids as applied to pathway 1:

Table 11.1 U.S.EPA Risk Assessment Pathways for Application of Biosolids to Soil

Pathway	Highly Exposed Individual (HEI) or Receptor
1. Sludge–soil– plant–human	Protection of consumers who eat produce grown in soil using sewage sludge. 2.5% of intake of grains, vegetables, potatoes, legumes and garden fruit is assumed to be grown on sludge-enriched soil.
2. Sludge–soil– plant–home gardener	Home gardener who produces and consumes potatoes, leafy vegetables, legume vegetables, root vegetables and garden fruit. 60% of HEI's diet is assumed to be grown on sludge-amended soil.
3. Sludge–soil–child	Assessment of the hazard to a child ingesting undiluted sewage sludge. Sewage sludge ingestion was 0.2 g dry weight/day/5 years.
4. Sludge–soil– plant–animal–human	Human exposure from consumption of animal products. 40% of the HEI's diet of meat, dairy products, or eggs is assumed to come from animals consuming feed from soil to which sludge was applied. In a nonagricultural setting, a human consumes products from wild animals that ate plants grown on sludge-amended soil. The HEI is also assumed to be exposed to a background intake of a pollutant.
5. Sludge–soil– animal–human	The direct injection of sewage sludge by animals and the consumption by humans of the contaminated tissue. Direct ingestion of sludge by animals, where it has been surface applied. When sewage sludge is injected into the soil or mixed into the plow layer, grazing animals ingest the soil containing sludge. The HEI is also assumed to be exposed to a background intake of a pollutant.
6. Sludge–soil– plant–animal toxicity	Protection of the highly sensitive/exposed herbivorous livestock that consume plants grown on sewage sludge-amended soil. It is assumed that the livestock diet consists of 100% forage grown on sewage sludge-amended land and that the animal is exposed to a background pollutant intake.
7. Sludge–soil– animal toxicity	Protection of the highly sensitive/highly exposed herbivorous livestock which incidentally consume sewage sludge adhering to forage crops and/or sewage sludge on the soil surface. The amount of sewage sludge in the livestock diet is assumed to be 1.5% and the animal is exposed to a background pollutant intake.
8. Sludge–soil–plant toxicity	Evaluation of risk to plant growth (phytotoxicity) from pollutants in sludge. Probability of 50% reduction of plant growth associated with a low probability of 1×10^{-4}.
9. Sludge–soil– soil–biota toxicity	Protection of highly exposed/highly sensitive soil biota. Criteria for this pathway have been set using earthworm (*Eisenia foetida*) data.
10. Sludge–soil–soil biota–predator of soil biota toxicity	Protection of the highly sensitive/highly exposed soil biota predator. Sensitive wildlife that consume soil biota that has been feeding on sewage sludge-amended soil. Chronic exposure assumes that 33% of the sensitive species' diet is soil biota.
11. Sludge–soil– airborne dust– human	Tractor operator exposed to 10 mg/m³ total dust while tilling a field to which sewage sludge has been applied.
12. Sludge–soil– surface water– contaminated water–fish toxicity–human toxicity.	Protection of human health and aquatic life. Risk to surface water associated with run-off of pollutants from soil on which sewage sludge has been applied. Water quality criteria are designed to protect human health assuming exposure through consumption of drinking water and resident fish and to protect aquatic life.
13. Sludge–soil–air– human	Protection of members of farm households inhaling vapors of any volatile pollutant that may be in the sewage sludge when it is applied to the land. This pathway is not applicable to inorganic pollutants. It is assumed that the total amount of pollutant spread in each year would be vaporized during that year.
14. Sludge–soil– groundwater– human	Exposure of individuals drinking water from groundwater directly below a field to which sewage sludge has been applied.

$$RIA = \frac{0.0008*70}{1.0} 0.012*10^3 = 44\,\text{mg} \qquad (11.2)$$

The RIA is used to determine the cumulative amount of a pollutant that can be land applied from biosolids for the selected pathway without adverse effects. In this case, Pathway 1 (an adult over a lifetime, consumes crops grown on biosolids-amended soil) is used as an illustration.

$$RP_c = \frac{RIA}{\hat{A}(UC \times DC \times FC} \qquad (11.3)$$

where
RP$_c$ = the cumulative amount of a pollutant that can be land applied, without adverse effects, from biosolids exposure through the pathway evaluated.
UC   = plant uptake slope for pollutant from soil amended with biosolids.
DC   = dietary consumption of different food groups grown in soils amended with biosolids.
FC   = fraction of different food groups assumed to be grown in soils amended with biosolids.

The product of UC $\times$ DC $\times$ FC is 0.00654. Therefore, for arsenic in biosolids as applied to Pathway 1, the cumulative amount that can be land applied without adverse effects is 6700 kg/ha of As biosolids.

$$RP_c = \frac{44}{\hat{A}UC \times DC \times FC} = 6700\,\text{kg/ha} \qquad (11.4)$$

The most limiting pathway for As was Pathway 3, a child ingesting biosolids. This analysis is shown below:

$$RIA = \frac{0.0008*16}{1.0} 0.0045 = 8.3\,\text{mg} \qquad (11.5)$$

The principal difference in the calculation of equation (5) vs. equation (2) is the body weight (BW) of a child (16 kg) vs. that of an adult (70 kg). Also, the total intake of As for a child is 0.0045 mg/day vs. 0.012 mg/day for the adult.

The next step in calculating the concentration of a pollutant (RSC) in biosolids that can be expected not to produce adverse effects is as follows:

$$RSC = \frac{RIA}{I_s*DE} \qquad (11.6)$$

where

RSC = concentration of a pollutant in biosolids that can be ingested without the
         expectation of adverse effects
RIA = amount of pollutant ingested by humans without expectation of adverse effects
I_s = rate of biosolids ingestion by children
DE   = exposure duration adjustment; an attempt to consider less than lifetime exposure

The RSC for As concentration in biosolids ingested by children is calculated as follows:

$$RSC = \frac{8.3}{0.2*1} = 41\,mg \qquad (11.7)$$

Similar assessments were conducted for other potential As toxicity pathways.

Phytotoxicity of inorganic elements (Pathway 8) was evaluated in two different methods:

Method I

- A phytotoxicity threshold (PT_{50}) was established. This value is the concentration of a pollutant that can cause a 50% reduction in plant growth. This was based on short-term data.
- A calculation was made to determine the probability that the heavy metal concentration in plants grown on biosolids-amended soil would exceed the PT_{50} at various metal loadings using field studies.
- An acceptable level of tolerable risk exceeding the PT_{50} was set at 0.01 (i.e., 1 out of 100 times).
- The allowable loading rate of biosolids (RP) was the rate that would have less than 0.01 probability of causing the PT_{50} to be exceeded.

The example provided below is for zinc.

- The PT_{50} for Zn = 1975 µg Zn/g plant tissue dry weight.
- The probability that corn grown on biosolids-amended soils would exceed the PT_{50} was computed for 12 Zn loading ranges.
- The tolerable risk for exceeding PT_{50} was set at 0.01.
- None of the loading rates evaluated exceeded the probability of 0.01. Therefore the highest loading rate evaluated (3,500 kg Zn/ha) was chosen as the allowable loading rate (RP) for biosolids that would not cause a significant phytotoxic effect in corn. RP = 3500 kg Zn/ha.

Method II

$$RP = \frac{TPCBC}{UC} \qquad (11.8)$$

This method evaluated the lowest-observed-adverse-effects-level (LOAEL). The reference cumulative application rate of a (RP) of Zn was calculated as follows.

Where:

RP   = The amount of a pollutant that can be applied to a hectare of land without
        expectation of adverse effects
TPC = The concentration of a pollutant in a sensitive plant tissue species (e.g., lettuce,
        as opposed to a less sensitive species, such as corn, used in method I)
BC   = Background concentration of pollutant in plant tissue
UC   = Plant uptake of pollutant from soil/biosolids

For Zn the following parameters were used:

TPC = 400 mg of Zn/g plant tissue in lettuce dry weight (mg/g DW)
BC   = 47.0 mg of Zn/g plant tissue of lettuce DW
UC   = 0.125 mg of Zn/g of lettuce plant tissue (kg of Zn per ha) (mg/g DW)(kg/ha)

The calculation of RP for Zn is as follows:

$$RP = \frac{40047.0}{0}.125 = 2800\,kgZn\,/\,ha$$

A comparison of the results of Method I (3500 kg Zn/ha) and II (2800 kg Zn/ha) shows that the more restrictive result was an RP of 2800 kg Zn/ha. The limit set for Pathway 8 was the pollutant limit used in the Part 503 rule for Zn.

Table 11.2 summarizes the pollutant limits for heavy metals in biosolids and biosolid products (USEPA, 1995). Prior to reviewing the pollutant limits, an explanation of the following definitions is in order:

- Ceiling concentration – This is the maximum concentration in mg/kg of an inorganic pollutant (heavy metal) in biosolids compost that is allowed for land application. If biosolids contain pollutants above these levels, the product may not be applied to land. Below this limit, other criteria may restrict its use. States may issue regulations that have lower limits, but not higher ones.
- Pollutant concentration (PC) limits – The pollutant concentration limit is the maximum concentration in mg/kg of an inorganic pollutant and applies to Class B biosolids.
- Cumulative Pollutant Loading Rate (CPLR) – This is the maximum amount of an inorganic pollutant that can be applied to an area of land.
- Alternative Pollutant Limit (APL) – This is the highest level of a given heavy metal in biosolids that is permitted in materials to be marketed.
- Exceptional Quality Biosolids (EQ) – Although this term is not used specifically in the 503 regulations, it is used in documents published by USEPA explaining the 503 regulations (USEPA, 1994). It refers to the concentration of a low pollutant in biosolids that meets the USEPA no observed adverse effects limits (NOAEL) criteria, as well as the pathogen and vector attraction reduction requirements.
- Annual Pollutant Loading Rate (APLR) – This is the highest annual (365 days) rate of application of each pollutant to land in kg/ha.

Table 11.2 Pollutant Limits for Heavy Metals in Biosolids and Biosolids Products

Pollutant	Ceiling Concentration Limits for all Biosolids Applied to Land mg/kg[1]	Pollutant Concentration Limits for EQ and PC Biosolids mg/kg[1]	Cumulative Pollutant Loading Rate Limits for CPLR Biosolids kg/ha	Annual Pollutant Loading Rate Limits for APLR Biosolids kg/ha/365-Day Period
Arsenic	75	41	41	2.0
Cadmium	85	39	39	1.9
Copper	4,300	1,500	1,500	75
Lead	840	300	300	15
Mercury	57	17	17	0.85
Molybdenum[2]	75	–	–	–
Nickel	420	420	420	21
Selenium	100	36	100	5.0
Zinc	7,500	2,800	2,800	140
Applies to:	All biosolids that are land applied	Bulk biosolids and bagged biosolids[3]	Bulk biosolids	Bagged biosolids[3]
From Part 503	Table 1, Section 503.13	Table 3, Section 503.13	Table 2, Section 503.13	Table 4, Section 503.13

[1] Dry-weight basis.

[2] The limits for molybdenum were deleted from the 503 rule on February 25, 1994 (*Fed. Reg.*, Vol. 39, No. 38, p. 9095).

[3] Bagged biosolids sold or given away in bag or other container.

[4] Chromium deleted from regulations and selenium modified in 1995.

Source: USEPA, 1995.

In addition to pollutant limits, the 503 rule also required pathogen and vector attraction reduction criteria. The basis for the 503 pathogen requirements are provided in the USEPA document Technical Support Document for Reduction of Pathogens and Vector Attraction in Sewage Sludge (USEPA, 1992). In October 1999, USEPA issued a revision of the document Environmental Regulations and Technology Control of Pathogens and Vector Attraction in Sewage Sludge (EPA/625/R-92-013). In the previous USEPA 257 regulations, the only requirements for composting were based on time-temperature relationships.

In a 1988 study, Yanko demonstrated that regrowth of pathogens occurs in biosolids compost. In this study, salmonellae were detected 165 times in 365 measurements. No salmonellae were detected in the 86 measurements for which the fecal coliform densities were less than 1000 MPN (most probable number)/g. This indicated that the potential for finding salmonellae would be highly unlikely when the fecal coliform densities were less than 1000 MPN/g (USEPA, 1992; Farrell, 1992). USEPA (1992) states that the reason for alternately using the fecal coliform test or the salmonellae test is that fecal coliform can regrow to levels exceeding 1000 MPN/g, but once totally eliminated, salmonellae can never grow.

USEPA required that sewage sludge be treated to reduce the potential for insects, birds, rodents and domestic animals to transport sewage sludge and pathogens to humans. These vectors are attraced to sewage sludge as a food source. The Part 503 regulations required that sewage sludge be treated to reduce vector attraction. Vector attraction reduction (VAR) could be achieved in two ways: 1) by treating the sewage sludge to the point at which vectors will no longer be attracted to it and 2) by placing a barrier between the sewage sludge and vectors.

Subpart D (503.33) provides for 12 options necessary to demonstrate VAR. Options 1 to 8 apply to sewage sludge that has been treated to reduce vector attraction. These options consist of either operating conditions or tests to demonstrate the vector attraction has been reduced in the treated sewage sludge. Options 9 through 11 are "barrier" options. These aply to both biosolids and domestic septage. Details are provided in the USEPA document Control of Pathogens and Vector Attraction in Sewage Sludge, EPA/625/R-92/013, 1999.

For Class A biosolids, pathogen reduction must take place before or at the same time as vector attraction reduction occurs. In the options 6, 7 and 8, this does not apply.

The 503 regulations also provide sampling and analysis methodologies. One of the most important aspects of the 503 regulations, which impact land application methodologies, is liability. Direct land application, whether by a public or private entity, is the legal responsibility of the producer of biosolids. If a municipality or its contractor violates the permit requirement for land application of biosolids, the producer, its employees and the contractor are subject to civil and criminal action. For example, if a contractor violates the municipality's permit to apply a specific quantity of biosolids containing the 503 heavy metal limitations, the contractor, the municipality and any knowledgeable individuals can be liable and sued for both criminal and civil damages.

The distribution and marketing of biosolids products, such as compost, does not entail similar liability. A contractor or individual purchasing compost containing the limit of heavy metals and distributing or marketing the compost at excessive rates does not face criminal or civil charges. Only product liability litigation could result (e.g., if compost is provided to a user without adequate instruction on its use and it causes phytotoxicity).

Pathogens in sewage sludge, biosolids and septage are regulated under Subpart D of the part 503 rule (USEPA, 1999). Two classes are designated: Class A and Class B. Class A is designed so that pathogens are not detected in biosolids or biosolids products. These include bulk or bagged products that are given away for home gardens or other horticultural uses. Once sewage sludge is treated to meet Class A or Class B, it can be designated as biosolids. This distinguishes it from untreated material.

The pathogen regulations involve three aspects:

1. Specific pathogen requirements
2. Process requirements
3. Vector attraction requirements (VAR)

CLASS A REQUIREMENTS

Pathogen requirements are as follows:

- The density of fecal coliform in the sewage sludge or biosolids must be less than 1,000 most probable number (MPN)/g total solids (dry-weight basis).

or

- The density of *Salmonella* sp. bacteria in the sewage sludge or biosolids must be less than 3 MPN per 4 g of total solids (dry-weight basis).

Either of these requirements must be met at one of the following times:

- When sewage sludge or biosolids are used or disposed
- When sewage sludge or biosolids products such as compost, alkaline stabilized material, or heat-dried products are prepared for sale or giveaway in a bag or other container for land application; or
- When the sewage sludge or biosolids product or derived materials are prepared to meet the requirements for EQ biosolids.

Pathogen reduction must take place before or at the same time as VAR.

Process Requirements

Six alternatives exist for treating sewage sludge or biosolids so that they can meet Class A pathogen requirements. These are:

Alternative 1—Thermally Treated Sewage Sludge [(503.32(a)(3)]

Pathogen requirements as stated above must be met.

Biosolids must be subjected to one of four time-temperature regimes. Each regime is based on the percentage of solids of the sewage sludge and on operating parameters. Vector attraction (VAR) is met by reducing the volatile solids by more than 38%.

Alternative 2—Sewage Sludge Treated in a High pH–Temperature Process (Alkaline Treatment) [503.329(a)(94)].

The process conditions required are:

- Elevated pH to greater than 12 hours and maintaining the pH for more than 72 hours.
- Maintaining the temperature above 52°C (126°F) throughout the sewage sludge for at least 12 hours during the period that the pH is greater than 12.
- Air drying to over 50% solids after the 72-hour period of elevated pH.
- VAR option 6, pH adjustment; pH to remain elevated until use/disposal.

Alternative 3—Sewage Sludge Treated in Other Processes [503.32(a)(5)]

This alternative relies on comprehensive monitoring of bacteria, enteric viruses and viable helminth ova to demonstrate adequate reduction of pathogens. VAR depends on the process by which pathogen reduction is met (one to 11 VAR options).

Alternative 4—Sewage Sludge Treated in Unknown Processes [503.31(a)(6)].

The requirements are similar to Alternative 3. Pathogen monitoring is required and testing is to be done at the time the sewage sludge is used or disposed. VAR is demonstrated by showing a 38% volatile solids reduction. VAR depends on the process by which pathogen reduction is met (one To 11 VAR options).

Alternative 5—Use of Process to Further Reduce Pathogens (PFRP) [503.32(a)(7)]

Sewage sludge is considered Class A if is treated by PFRP as listed and the Class A pathogen requirements are met. VAR must be met. PFRP processes include:

- Composting
- Heat drying
- Heat treatment
- Beta ray irradiation
- Gamma ray irradiation
- Pasteurization

Alternative 6—Use of a Process Equivalent to PFRP [503.32(a)(8)]

The requirements are similar to Alternative 5.

CLASS B REQUIREMENTS

Class B requirements can be met in three different ways.

- Monitoring of indicator organisms — tests for fecal coliform density as an indicator organism for all pathogens. The geometric mean of seven samples shall be less than two million MPNs per gram total solids or less than two million CFUs (colony-forming unit) per gram of total solids at the time of use or disposal.
- Sewage sludge is treated in processes to significantly reduce pathogens (PSRP).
- Sewage sludge treated in a process equivalent to a PSRP.

In addition to these requirements, site restrictions for Class B sewage sludge are applied to land.

Table 11.3 Maximum Acceptable Heavy Metal Concentrations and Maximum Acceptable Cumulative Heavy Metal Additions to Soil

Heavy Metal	Maximum Acceptable Metal Concentrations mg/kg[1]	Maximum Acceptable Cumulative Heavy Metal Additions to Soil kg/ha
As	75	15
Cd	20	4
Cu	150	30
Pb	500	100
Hg	5	1
Mo	20	4
Ni	180	36
Se	14	2.8
Zn	1,850	370

[1] In processed sewage, sewage-based products and other by-products containing 5% N or less and represented for sale as fertilizer.
Source: Agriculture and AgriFood Canada, Trade Memorandum, T-4-93, July 1995.

CANADA

Canadian provinces set regulations or guidelines for biosolids' heavy metal concentrations and limits to be applied to soils. Table 11.3 shows the maximum acceptable heavy metal concentrations and maximum acceptable heavy metal additions to soil as recommended by the Ministry of Agriculture. Canadian provinces can set their own limits, which are shown in Table 11.4.

Table 11.4 Canadian Provincial Guidelines for Heavy Metals

Heavy Metal	Nova Scotia	Ontario	Alberta Class A	British Columbia
		Concentration - mg/kg dw		
Cd	2.6	3	<2	2.6
Cu	100	60	<80	100
Cr	210	50	<100	210
Pb	150	150	<50	150
Hg	0.83	0.15	<0.2	0.83
Ni	50	60	<32	50
Zn	315	500	<120	315

Source: Dillon, 1994.

A recent document prepared for the Water Environment Federation of Ontario (WEAO, 2001) reviewed the literature on fate and significance of selected contaminants in sewage biosolids applied to agricultural land. The Ontario guidelines, introduced in the mid-1970s, recommended applying liquid anaerobically stabilized biosolids at a rate not to exceed 135 kg of plant available N per hectare per 5 years. This is based on the amount of N needed to grow corn. The heavy metal guidelines,

Table 11.5 Ontario Guidelines for Heavy Metals in Biosolids and Agricultural Soils

Heavy Metal	Anaerobic Biosolids Minimum (Ammonium + Nitrate) Nitrogen to Metal Ratios	Aerobic, Dewatered and Dried Biosolidsmg/kg Dry Weight	Maximum Permissible Metal Concentrations in Soil mg/kg Dry Weight	Maximum Permissible Metal Loading to Soil mg/kg Dry Weight
Arsenic	100	170	14	14
Cadmium	500	34	1.6	1.6
Cobalt	50	340	20	30
Chromium	6	2800	120	210
Copper	10	1700	100	150
Lead	15	1100	60	90
Mercury	1500	11	0.5	0.8
Molybdenum	180	94	4	4
Nickel	40	420	32	32
Selenium	500	34	1.6	2.4
Zinc	4	4200	220	330

Based on Guidelines for the Utilization of Biosolids and Other Wastes on Agricultural Land. Ontario Ministry of the Environment and Energy and Ontario Ministry of Agriculture, Food and Rural Affairs, Toronto, ON.
Source: WEAO, 2001. With permission.

based on background levels for soils, are shown in Table 11.5. This concept is flawed because use of pesticides, herbicides and fertilizers containing heavy metals is allowed, even though they exceed background levels.

EUROPE

Regulations vary among European countries. These regulations are essentially based on the no-net-degradation approach that has strict limitations. As McGrath et al. (1994) point out, the scientific basis for many of the limits in European countries is difficult to find. Furthermore, in those countries that have a short history of biosolids disposal, the scientific basis is totally lacking. The European Union (EU) has its own standards. The EU is a federation of 15 sovereign states in Western Europe. While the EU encourages the use of biosolids, the EU directive 86/278/EEC regulates its use to prevent harm to the environment — especially soils (Langenkamp and Part, 2001). Table 11.6 shows the current and projected heavy metal limits proposed by the EU (Matthews, 1999; EC, 2000).

Differences in several European countries are illustrated in Table 11.7. The heterogeneity in ecological and economic conditions makes it very difficult to establish consensus in Europe. In some cases, ecology or the need to preserve the limited agricultural soils in a country is the driving force. Many soils in central Europe are contaminated due to long-term industrial pollution. In other countries, economics is the driving force. Farmers need biosolids as a nutrient source and to reduce their fertilizer costs. It appears that the EU standards set an overall condition

Table 11.6 Limit Values for Concentrations of Heavy Metals in Biosolids and Limits to Annual Additions of Metals to Soil Based on a 10-Year Average

Element	Current		Proposed 2015		Proposed 2025	
	Limit Values for Concentrations mg/kg dm	Limit Values for Annual Additions g/ha/y	Limit Values for Concentrations mg/kg dm	Limit Values for Annual Additions g/ha/y	Limit Values for Concentrations mg/kg dm	Limit Values for Annual Additions g/ha/y
Cd	20–40	150	5	15	2	6
Cr	–	–	800	2400	600	1800
Cu	1000–1750	12000	800	2400	600	1800
Hg	16–25	100	5	15	2	6
Ni	300–400	3000	200	600	100	300
Pb	750–1200	1500	1500	1500	200	600
Zn	2500–4000	30000	2000	6000	1500	4500

Source: McGrath et al., 1994.

Table 11.7 Comparison of Concentration Limits for Heavy Metals in Biosolids Used for Land Application between the United States and Several European Countries

Element	U.S.[1]	EU	Netherlands	Sweden	Denmark	Germany
			mg/kg/Dry Weight			
As	41	—	0.15	—	25	—
Cd	39	20–40	1.25	2	0.8	5–10
Cr	(1200)[2]	—	75	100	100	900
Cu	1500	1000–1750	75	600	1000	800
Hg	17	16–25	0.75	2.5	0.8	8
Pb	300	750–1200	100	100	60–120	900
Ni	420	300–400	30	50	30	200
Zn	2800	2500–4000	300	800	4000	2000–2500

[1] USEPA regulations 40 CFR 503.
[2] Cr subsequently deleted from regulations.

for the countries, yet individual countries, such as Sweden, can have more restrictive standards.

Table 11.7 shows that considerable variation exists within the EU. This variation likely reflects the lack of scientific basis. It appears that, in many cases, values were arbitrarily selected and not based on risk assessment. Harrison et al. (1999) indicate that countries such as Sweden, Denmark and the Netherlands use a philosophy of "do no harm" to protect soil quality. This concept is based on the "no net degradation" concept (i.e., that inputs do not exceed outputs).

Loading limits are shown in Table 11.8.

Table 11.8 A Comparison of Loading Limits in kg/ha/Year in the United States and European Union[1]

Element	USEPA	EU[2]	Lowest member state	Proposed future EU limits Current to 2015	2015–25	After 2025
As	2	–	0.25 Denmark[3]	–	–	–
Cd	1.9	0.15	0.002 Sweden	0.03	0.015	0.006
Cr	150	–	0.1 Sweden	3.0	2.4	1.8
Cu	75	12	0.6 Sweden	3.0	2.4	4.8
Hg	0.85	0.1	0.001 Finland	0.03	0.015	0.006
Pb	15	15	0.1 Sweden	2.25	1.5	0.6
Ni	21	3	0.05 Sweden	0.9	0.6	0.3
Se	5	–	0.15 UK[4]	–	–	–
Zn	140	30	0.8 Sweden	7.5	6.0	4.5

[1] Excluding the Netherlands.
[2] The loading rate is averaged over 10 years.
[3] Only the U.K. and Denmark regulate As.
[4] Only the U.K. regulates Se.
Source: Evens, 2001.

Several countries also have standards for concentrations of organic contaminants in biosolids. EU standards are as follows: Halogenated organic compounds 500 mg/kg dm; DEHP, 100 mg/kg dm; LAS, 2600 mg/kg dm; NP/NPE 50 mg/kg dm; PAH, 6 mg/kg dm; PCB, 0.8 mg/kg dm; and PCDD/F, 100 mg TEQ/kg dm. Denmark has standards for DEHP, LAS, NP/NPE and PAH; Sweden for NP/NPE, PAH and PCB; and Germany for PCB and PCDD/F. In these countries, the standards are equal to or lower than the EU standards.

CONCLUSION

The USEPA regulations are risk based. The agency is in the process of reviewing and supplementing the regulations. The next round of regulations will address selective organic compounds.

Although the scientific community in the United States in general has supported the USEPA regulations, a vocal minority believes that the 503 rule is not sufficiently protective of human health and the environment. Their concerns merit a brief discussion.

Several points need to be made or reemphasized. The 503 rule was risk based. This rule is constantly subject to changes as new research data become available. No other waste or product has been subjected to such extensive research and examination. In 1996, the National Research Council of the National Academy of Sciences issued a report entitled, "Uses of Reclaimed Water and Sludge in Food Crop Production." A second review by the National Research Council was published in draft form on July 2, 2002.

Here are some of the major findings and recommendations:

1. The committee recognized that land application of biosolids is a widely used, practical option for managing the large volume of sewage sludge generated in the United States that otherwise would be incinerated or landfilled.
2. The committee did not find documented scientific evidence that the Part 503 regulations had failed to protect public health. Additional scientific work is needed to reduce uncertainties about the potential adverse public effects from exposure to biosolids.
3. There is a need to update the scientific basis of the rule as related to chemical and pathogen standards, demonstrate that there is effective enforcment of the Part 503 rule and validate the effectiveness of biosolids management practices.
   - There is a lack of exposure and health information on poulations exposed to biosolids.
   - There has been no substantial reassessment to determine whether the chemical and pathogen standards promulgated under the Part 503 Rule in 1993 are supported by current scientific data and risk assessment.
   - The technical basis of the 1993 chemical standards for biosolids is outdated.
4. The committee supports USEPA's approach to establishing pathogen reduction requirements and monitoring indicator organisms. Better documentation is needed using current pathogen detection methods; there is a need for better research on environmental persistence and for determining dose–response relationships to verify that the current management controls for pathogens are adequate to maintain minimal concentrations of pathogens over extended periods of time.
5. The major recommendations are:
   - Use improved risk-assessment methods to establish standards for chemicals and pathogens.
   - Conduct a national survey of chemicals and pathogens in sewage sludge.
   - Establish a procedural framework for implementing human health investigations.
   - Increase the resources for USEPA's biosolids program.

For more information, a prepublication of the report, entitled Biosolids Applied to Land: Advancing Standards and Practices, is available from the Committee on Toxicants and Pathogens in Biosolids Applied to Land, Board on Environmental Studies and Toxicology, Division of Earth and Life Sciences, National Research Council, Washington, D.C.

Several scientists at Cornell University and at the Cornell University Waste Management Institute, Center for the Environment, indicated that the current U.S. federal regulations governing land application of biosolids do not appear to adequately protect human health and the environment (Harrison et al., 1999). They point out that the U.S. regulations are much less protective than the Canadian or European guidelines or regulations. Their main arguments follow:

- Pollution is allowed to reach maximum "acceptable" level. Without a very good understanding of pathways and processes, it is unwise to allow pollutants to reach calculated maximum acceptable values.
- Each exposure pathway was evaluated separately and did not account for multiple pathways of exposure or synergy. Often risk assessments use an additive approach.
- USEPA calculated cancer risk of 1 in 10,000 vs. 1 in 1,000,000. This resulted in less restrictive values.

- The soil ingestion rate by children of 200 mg/day may be too low and the period of five years too short.
- The pollutant intake through food was underestimated. The USDA recommended intake of vegetables and fruits is much higher than the intake estimates used by USEPA.
- The plant uptake coefficients were very low. These coefficients express the amount an element is taken up by the plant as compared with the amount in the soil.
- Many pollutants are not regulated or monitored. These include pollutants where insufficient data exist, such as synthetic organic chemicals and radioactivity in sludges. USEPA is currently evaluating organic chemicals and radioactivity in sludges.
- Ground and surface water calculations assume large dilution/attenuation.
- Insufficient attention was given to phytotoxic effects and effects on soil microorganisms and animals.

Regulations on biosolids management are important. The risk-based approach used by USEPA provides opportunity to modify regulations as better scientific data become available. The enforcment of the regulations is problematic. USEPA does not have the means to oversee the regulations. Even if states receive authorization to enforce regulations, they often lack the funds and personnel. Public confidence will be enhanced if the public perceives that the regulations are adequate to protect human health and the environment and that the regulations are enforced.

REFERENCES

Dillon, 1994, Compost Quality Objectives Study. Final Report to Alberta Environmental Protection, Alberta Agriculture, Alberta Health and Edmonton Board of Health. Prepared by GCG Dillon, Canada and E&A Environmental Consultants, Inc., Canton, MA.

EC, 2000, Working document on sludge, 3rd draft. April 27, 2000, EEC, Brussels, Belgium.

Evens, T., 2001, An update on developments in regulations affecting biosolids in the European Union, *WEF/AWWA Joint Residuals and Biosolids Management Conference. Biosolids 2001: Building Public Support,* Water Environment Federation, Alexandria, VA.

Farrell, J.B., 1992, Fecal pathogen control during composting, pp. 282–300, H.A.J. Hoitink and H.M. Keener (Eds.), *Science and Engineering of Composting: Design, Environmental, Microbiological and Utilization Aspects,* Renaissance, Worthington, OH.

Harrison, E.Z., M.B. McBride and D.R. Bouldin, 1999, Land application of sewage sludges: An appraisal of the U.S. regulations, *Int. J. Environ. Pollut.* 11(1): 1–36.

Kennedy, J., 1992, A review of composting criteria, *Proc. of the Composting Council of Canada, Second Annual Meeting,* Ottawa.

Langenkamp, H. and P. Part, 2001, Organic contaminants in sewage sludge for agricultural use, European Commission Joint Research Centre Institute for Environment and Sustainability, Soil and Waste Unit.

Matthews, P., 1999, Sewage sludge treatment and biosolids management in Europe, *Sewage Sludge Treatment and Disposal in Spain,* IQPC, Ltd., England; Madrid, Spain.

McGrath, S.P., M.C. Chang, A.L. Page and E. Witter, 1994, Land application of sewage sludge: Scientific perspectives of heavy metal loading limits in Europe and the United States, *Environ. Rev.* 2 :108–118.

NRC, 1996, *Use of Reclaimed Water and Sludge in Food Crop Production, Water Science and Technology Board/ National Research Council,* National Academy of Sciences, National Academy Press, Washington, D.C.

Ryan, J.A. and R.L. Chaney, 1995, Issues of risk assessment and its utility in development of soil standards: The 503 methodology as an example, *Proc. 3rd Int. Symp. Biogeochemistry Trace Elements,* Paris.

USEPA, 1992, Technical support document for reduction of pathogens and vector attraction in sewage sludge, EPA 822/R–93–004, U.S. Environmental Protection Agency, Washington, D.C.

USEPA, 1994, A plain English guide to the EPA Part 503 biosolids rule, EPA/832/R–93/003, U.S. Environmental Protection Agency, Washington, D.C.

USEPA, 1995, Standards for the use or disposal of sewage sludge; final rule and proposed rule, U.S. Environmental Protection Agency, *Fed. Reg.* Vol. 60, No. 206, Washington, D.C.

USEPA, 1999, Control of pathogens and vector attraction in sewage sludge, U.S. Environmental Protection Agency, Office of Research and Development, National Risk Management Research Laboratory, Center for Environmental Research Information, EPA/625/R-92-013, Cincinnati, OH.

WEAO, 2001, Fate and significance of selected contaminants in sewage biosolids applied to agricultural land through literature review and consultation with stakeholder groups. Final Report. Prepared by R.V. Anderson Assoc., Water Environment Association of Ontario.

Yanko, W.A.,1988, Occurrence of pathogens in distribution and marketing municipal sludges. Rep. No. EPA–6/1–87–014. (NTIS PB 88-154273/AS.) Cincinnati Health Effects Research Laboratory, Cincinnati, OH.

Index

A

Acidity, 22–23, 57, 58
Actinomycetes, 121
Activated sludge, 145–146, *see also* Agricultural
    crops; Food crops; Sewage
    sludge/biosolids
Adsorption, viruses, 135
Aeration, 73
Aerobic digestion, 6, 22, 10, 116
Aerosolized pathogens, 129, *see also* Pathogens
AGRICOLA, 144
Agricultural crops, land application of sewage
    sludge/biosolids
    forestry and reclamation, 148–152
    research results prior to 1970, 145–146
    research results 1970 to 2001, 146–148
Agricultural soils, 43, *see also* Soils
Alcaligenes spp., 33
Alfalfa, 146
Alkaline stabilization, 6, 10, 34, 121
Alkaline treatment, 184
Alkylbenzene sulphonates, 19, 20, 21
Alkylphenol ethoxylates, 88
Alkylphenols, 19, 20, 21, 22
Alternative pollutant limit (APL), 181
Ammonia, 33–34
Ammonification, 32, 34–35
Anaerobic bacteria, 33, *see also* Individual entries
Anaerobic digestion, 6, 22, 10, 116–118, *see also*
    Mesophilic anaerobic digestion
Analysis methodologies, 183, *see also* 503 Rule
Animals
    arsenic content, 47
    cadmium content, 48–51
    chromium content, 52
    copper content, 53–54
    ingestion of plants/soils and contamination of
        food chain, 94

land application of biosolids, effects
    domestic, 160–164
    earthworms, 167–168
    microbes, 166–167
    wildlife, 164–165
lead content, 55
mercury content, 56
molybdenum content, 57
nickel content, 59
routes of pathogen transmission, 130
Annual pollutant loading rate (APLR), 181
Anthracite refuse bank, 152, *see also* Reclamation
Antibiotics, 121
APL, *see* Alternative pollutant limit
APLR, *see* Annual pollutant loading rate
Aporrectodea tuberculata, 167, 168
Aquatic environment, 22, *see also* Environment
Arsenic, 47–48, 177, 179
Ascaris spp., 116, 118, 137, 138
ATAD, *see* Autothermal thermophilic aerobic
    digestion
Atmospheric deposition, 87, 90
Autothermal thermophilic aerobic digestion
    (ATAD), 116, 117

B

Bacillus spp., 121
Bacteria, *see also* Individual entries
    soil content, 131–134
        survival and soil conditions, 136
    wastewater treatment and efficiency removal,
        111, 113
Bacteriophages, 136
Barley, 146
Barrier options, 183
BCFs, *see* Bioconcentration factors
BDEs, *see* Brominated diphenyl ethers
Best achievable approach, 174

193

Milton Keynes UK
Ingram Content Group UK Ltd.
UKHW040058071024
449327UK00019B/639